SU Carburettors Owners Workshop Manual

by Don Peers

Types covered:
All SU variable choke carburettors designated as
Type H
Type HD
Type HS
Type HIF
and the Thermostatic starting carburettor (or 'Thermo')

ISBN 1 85010 059 4
ISBN 1 85010 018 7 (US)

Printed in England *(299 - 11M3)*

Haynes Publishing Group
Sparkford Nr Yeovil
Somerset BA22 7JJ England

Haynes Publications, Inc
861 Lawrence Drive
Newbury Park
California 91320 USA

Acknowledgements

We are indebted to SU Carburettors of Wood Lane, Erdington, Birmingham B24 9QS, part of British Leyland, for publishing their technical material.

Special thanks must go to A & N Parts of Vicarage Street, Yeovil, Somerset for helping us when requested, without a moment's hesitation, and for supplying us quickly with a sample HIF carburettor. John Haynes, our chairman, showed no sign of annoyance as we plundered some of the cars in his personal collection in order to look at the carburettors and we are most grateful to Brian Horsfall who used his dexterity to the full in 'stealing' other carburettors from visitors' cars, almost without their knowing, so that we could be sure of our facts.

Martin Penny helped Brian Horsfall take the carburettors to pieces whilst Les Brazier directed his lens at the procedures and provided us with the photographic sequences.

Terry Davey undertook all the drawings with his usual skill and integrity. Tim Parker guided the author, edited the text and put the book together.

About this manual

This is a manual written for the do-it-yourself car owners whose cars are fitted with SU carburettors. Because the carburettor is a fine precision instrument, the book has to assume a basic familiarity on the part of the reader. For example it assumes that the owner already knows how his car works, where the carburettor is fitted and how to set the contact breaker points, (if he doesn't then he can easily learn by reading the Haynes Owners Workshop Manual for his car) taking up the 'story' from that basic familiarity. It takes out the operation of the carburettor, isolating it from the rest of the car and explains its function and workings in full, only relating those to the others of the car when necessary. It is a complete, comprehensive manual for the SU carburettor in its own right.

A great deal of mystique surrounds the carburettor, its function and tuning; this manual explodes that mystique and guides the owner through the theory, maintenance, overhaul and tuning, using only the necessary tools which any reasonably equipped home workshop will possess. It does not over-complicate the issue by giving superfluous information which is of no practical use. By definition it is not possible to dictate when to look at your carburettor nor whether the law allows you to alter any setting on your car — in some countries environmental agencies specifically forbid it. It simply provides a guide and then the information on how to do it should you want to. Of course every effort is made to ensure that this book is complete and up-to-date. However, it should be realised that modifications in design and production are happening all the time, even in retrospect.

Contents

Safety first!

Professional motor mechanics are trained in safe working procedures. However enthusiastic you may be about getting on with the job in hand, do take the time to ensure that your safety is not put at risk. A moment's lack of attention can result in an accident, as can failure to observe certain elementary precautions.

There will always be new ways of having accidents, and the following points do not pretend to be a comprehensive list of all dangers; they are intended rather to make you aware of the risks and to encourage a safety-conscious approach to all work you carry out on your vehicle.

Essential DOs and DON'Ts

DON'T rely on a single jack when working underneath the vehicle. Always use reliable additional means of support, such as axle stands, securely placed under a part of the vehicle that you know will not give way.
DON'T attempt to loosen or tighten high-torque nuts (e.g. wheel hub nuts) while the vehicle is on a jack; it may be pulled off.
DON'T start the engine without first ascertaining that the transmission is in neutral (or 'Park' where applicable) and the parking brake applied.
DON'T suddenly remove the filler cap from a hot cooling system – cover it with a cloth and release the pressure gradually first, or you may get scalded by escaping coolant.
DON'T attempt to drain oil until you are sure it has cooled sufficiently to avoid scalding you.
DON'T grasp any part of the engine, exhaust or catalytic converter without first ascertaining that it is sufficiently cool to avoid burning you.
DON'T allow brake fluid or antifreeze to contact vehicle paintwork.
DON'T syphon toxic liquids such as fuel, brake fluid or antifreeze by mouth, or allow them to remain on your skin.
DON'T inhale dust – it may be injurious to health (see *Asbestos* below).
DON'T allow any spilt oil or grease to remain on the floor – wipe it up straight away, before someone slips on it.
DON'T use ill-fitting spanners or other tools which may slip and cause injury.
DON'T attempt to lift a heavy component which may be beyond your capability – get assistance.
DON'T rush to finish a job, or take unverified short cuts.
DON'T allow children or animals in or around an unattended vehicle.
DO wear eye protection when using power tools such as drill, sander, bench grinder etc, and when working under the vehicle.
DO use a barrier cream on your hands prior to undertaking dirty jobs – it will protect your skin from infection as well as making the dirt easier to remove afterwards; but make sure your hands aren't left slippery.
DO keep loose clothing (cuffs, tie etc) and long hair well out of the way of moving mechanical parts.
DO remove rings, wristwatch etc, before working on the vehicle – especially the electrical system.
DO ensure that any lifting tackle used has a safe working load rating adequate for the job.
DO keep your work area tidy – it is only too easy to fall over articles left lying around.
DO get someone to check periodically that all is well, when working alone on the vehicle.
DO carry out work in a logical sequence and check that everything is correctly assembled and tightened afterwards.
DO remember that your vehicle's safety affects that of yourself and others. If in doubt on any point, get specialist advice.
IF, in spite of following these precautions, you are unfortunate enough to injure yourself, seek medical attention as soon as possible.

Asbestos

Certain friction, insulating, sealing, and other products – such as brake linings, brake bands, clutch linings, torque converters, gaskets, etc – contain asbestos. *Extreme care must be taken to avoid inhalation of dust from such products since it is hazardous to health.* If in doubt, assume that they *do* contain asbestos.

Fire

Remember at all times that petrol (gasoline) is highly flammable. Never smoke, or have any kind of naked flame around, when working on the vehicle. But the risk does not end there – a spark caused by an electrical short-circuit, by two metal surfaces contacting each other, by careless use of tools, or even by static electricity built up in your body under certain conditions, can ignite petrol vapour, which in a confined space is highly explosive.

Always disconnect the battery earth (ground) terminal before working on any part of the fuel or electrical system, and never risk spilling fuel on to a hot engine or exhaust.

It is recommended that a fire extinguisher of a type suitable for fuel and electrical fires is kept handy in the garage or workplace at all times. Never try to extinguish a fuel or electrical fire with water.

Fumes

Certain fumes are highly toxic and can quickly cause unconsciousness and even death if inhaled to any extent. Petrol (gasoline) vapour comes into this category, as do the vapours from certain solvents such as trichloroethylene. Any draining or pouring of such volatile fluids should be done in a well ventilated area.

When using cleaning fluids and solvents, read the instructions carefully. Never use materials from unmarked containers – they may give off poisonous vapours.

Never run the engine of a motor vehicle in an enclosed space such as a garage. Exhaust fumes contain carbon monoxide which is extremely poisonous; if you need to run the engine, always do so in the open air or at least have the rear of the vehicle outside the workplace.

If you are fortunate enough to have the use of an inspection pit, never drain or pour petrol, and never run the engine, while the vehicle is standing over it; the fumes, being heavier than air, will concentrate in the pit with possibly lethal results.

The battery

Never cause a spark, or allow a naked light, near the vehicle's battery. It will normally be giving off a certain amount of hydrogen gas, which is highly explosive.

Always disconnect the battery earth (ground) terminal before working on the fuel or electrical systems.

If possible, loosen the filler plugs or cover when charging the battery from an external source. Do not charge at an excessive rate or the battery may burst.

Take care when topping up and when carrying the battery. The acid electrolyte, even when diluted, is very corrosive and should not be allowed to contact the eyes or skin.

If you ever need to prepare electrolyte yourself, always add the acid slowly to the water, and never the other way round. Protect against splashes by wearing rubber gloves and goggles.

When jump starting a car using a booster battery, for negative earth (ground) vehicles, connect the jump leads in the following sequence: First connect one jump lead between the positive (+) terminals of the two batteries. Then connect the other jump lead first to the negative (–) terminal of the booster battery, and then to a good earthing (ground) point on the vehicle to be started, at least 18 in (45 cm) from the battery if possible. Ensure that hands and jump leads are clear of any moving parts, and that the two vehicles do not touch. Disconnect the leads in the reverse order.

Mains electricity

When using an electric power tool, inspection light etc, which works from the mains, always ensure that the appliance is correctly connected to its plug and that, where necessary, it is properly earthed (grounded). Do not use such appliances in damp conditions and, again, beware of creating a spark or applying excessive heat in the vicinity of fuel or fuel vapour.

Ignition HT voltage

A severe electric shock can result from touching certain parts of the ignition system, such as the HT leads, when the engine is running or being cranked, particularly if components are damp or the insulation is defective. Where an electronic ignition system is fitted, the HT voltage is much higher and could prove fatal.

PART 1
Chapter 1 Fundamentals - theoretical

Contents

1 Function of a carburettor

A petrol engine, as fitted to a modern motor car, is an extremely complex mechanism and, for its satisfactory operation, must have the support and co-operation of equally complex ancillary equipment.

The two most important equipments are the ignition system for initiation of combustion of the fuel, and the carburation system, whose duty it is to ensure that the fuel is there in such quantities and qualities as may be required according to the needs of the moment.

The quantity is of course largely determined by the driver of the vehicle through the medium of his right foot actuating the 'accelerator' pedal, and is a prime factor contributing to the power developed by the engine.

The 'quality' is equally important but is rarely given a second thought by the driver, since it is generally not within his power to control, and even if it were, he could not hope to make such a good job of it as the carburettor, to whom this responsibility falls. The modern carburettor has almost entirely (there is one special condition – that of starting) relieved the driver of this burden, allowing him to devote his concentration and energy to 'higher things'.

The 'quality' control is the aspect of carburettor design to which great attention has been paid by designers in recent years, resulting in some very sophisticated instruments.

In order to understand fully the operation of a carburettor we must first understand the theoretical principles upon which the concept is based, and then apply them to the hardware.

First we examine the fuel.

2 Fuels

The fuel used in road-going vehicles is commonly referred to as pump petrol. It is obtained by distillation from *crude petroleum* which consists of a mixture of volatile compounds of hydrogen and carbon in liquid state, called *hydrocarbons.*

The hydrocarbons may be divided into three series (based on exact chemical composition) with several members in each, called *fractions.* The fractions exhibit similar properties, to a greater or lesser degree. The properties with which we are concerned are those that affect the performance of a petrol engine, in general terms, and are shown in Fig. 1.

Calorific Value (CV)

This is a measure of the heat energy contained per unit weight of the fuel when it is burnt, and is therefore indicative of its power producing capability. It is associated indirectly with the specific gravity (density) of the fuel, and although no strict relationship exists, generally the heavier the fuel (higher specific gravity) the greater the calorific value.

Volatility

This is a measure of the readiness of a fuel to evaporate. A highly volatile fuel evaporates easily at depressed temperatures and hence is of considerable value in starting an engine in cold weather. The volatility may be considered, in simple terms, to be proportional to the boiling point of the fuel, and high volatility is generally a characteristic of the lighter fractions.

Anti-detonation Value

This is a property of the fuel to resist detonation when fired in the cylinder of an engine of given compression ratio. Detonation is a phenomenon of combustion under pressure and is harmful to an engine. It is manifested by a characteristic audible knocking, commonly called 'pinking'.

From the foregoing it will be clear that the fuel used must be a blend of these fractions, since they each have advantages and disadvantages, and a compromise must be achieved to ensure suitability for universal use.

For instance, a fuel of high CV (paraffins) would give maximum power but would be impractical to use by itself due to the difficulty (perhaps impossibility) of starting (due to low volatility), and the limitation on the maximum compression ratio (about 4 : 1) which could be employed, in order to ensure freedom from destructive 'pinking'.

Similarly, a highly volatile fuel, which may seem ideally suited to cold climates, would be unsuitable for use by itself as it is a poor power producer.

The most important fact arising from these discussions on the component parts of the fuel is that *each fuel has a different requirement of oxygen for complete combustion* and, therefore, *any blend of these fractions will have a definite requirement of oxygen peculiar to that blend.* Any alteration of its composition will also affect the oxygen requirement.

The blend of these fractions which has been found to give the most satisfactory performance contains about 85% carbon and 15% hydrogen, (by weight), and requires about 15 pounds of air per pound of fuel for complete combustion. This ratio is termed the air/fuel (A/F) or mixture ratio, and it is this relationship which the carburettor must automatically maintain, or vary, as conditions dictate.

Fuel	Symbol	Relative Density	Boiling point, C	Air/fuel ratio by mass	Calorific value kj/kg
Paraffin Series					
Hexane	C_6H_{14}	0.663	68	15.2	
Heptane	C_7H_{16}	0.691	98	15.1	
Octane	C_8H_{18}	0.709	124	15.05	45.120
Nonane	C_9H_{20}	0.723	150	15	45.820
Decane	$C_{10}H_{22}$	0.735	172	15	
Undecane	$C_{11}H_{24}$	0.746	195	–	
Naphthene Series					
Cyclohexane	C_6H_{12}	0.780	81	14.7	
Hexahydro-toluene	C_7H_{14}	0.770	100	14.7	43.940 44.060
Hexahydro-xylene	C_8H_{16}	0.756	118	14.7	
Aromatic Series					
Benzene	C_6H_6	0.884	80	13.2	43.030
Toluene	C_7H_8	0.870	110	13.4	44.770
Xylene	C_8H_{10}	0.862	140	13.6	

Fig. 1. Analysis of crude petroleum

3 Effect of variation of A/F ratio on engine performance

It has been noted that the A/F ratio is 15 : 1 for perfect combustion. This is the ratio which results in no unburnt fuel and no excess air. It is termed the 'chemically correct' ratio. Petrol will, however, tolerate a wide range of mixture ratios, in practice from about 8 : 1 to 22 : 1.

The chemically correct ratio does not give maximum power or minimum fuel consumption. The ratios which give these optima are approximately 12 : 1 and 18 : 1 respectively, as shown in Fig. 2.

These curves are important with regard to the tuning of a carburettor as they enable the operator to tune the engine for maximum power or maximum economy, within defined limits.

The engine will run and give a mediocre performance, in respect of both power output and fuel consumption, outside these limits of 12 : 1 and 18 : 1. It is important to note that if the A/F ratio is 'leaned out' to say 20 : 1 that power decreases *and fuel consumption increases at the same time.*

Therefore, in an effort to achieve economy the reverse has in fact occurred. Similarly, if in an effort to extract more power, the A/F mixture is enriched beyond 12 : 1 the fuel consumption naturally rises *but the power output also decreases.*

In addition to power output and fuel consumption, the general effects of incorrect mixture ratios are many and varied.

At the extreme weak end of the scale, the speed of flame propagation within the combustion chamber is considerably slower. Consequently the charge is still burning when the piston has reached the end of its stroke and this leads to overheating and, in extreme cases, to burnt valves and burnt piston crowns. (Remember that a weak mixture will have an excess of air and this is then available for oxidation of internal metal parts).

If the mixture is too rich, carbon will rapidly form on the surface of the combustion chamber and piston crown, entailing a premature 'de-coke'. Spark plug performance will be impaired and, in extreme cases, dilution of lubricating oil on the cylinder walls due to unburnt petrol will cause accelerated wear.

With too rich or too lean a mixture, toxic emissions in the exhaust gases will increase. Although not of paramount importance at present — in Great Britain, at least — this is an aspect which is under strict control in some countries and has led to the design and introduction of 'emission control' engines and fuel systems.

It will now be appreciated that the mixture control is extremely critical if the engine is to operate satisfactorily.

From the foregoing it may be thought necessary to determine the A/F ratio when tuning the carburettor. This is not so, for if the carburettor is adjusted in a certain manner the A/F ratio is automatically correct within the limits stated. The A/F ratio can be measured, for verification or for fine adjustment, by analysing a sample of exhaust gas. As the sophisticated equipment necessary for this is not normally available to the amateur its method of use will not be discussed. Its use will, however, play an increasingly important part in carburettor tuning in future years as more government agencies become aware of the necessity to reduce toxic emissions. Our carburettors will then be tuned on a different criterion of performance, that of minimum health hazard, and the amateur will then no longer be permitted to tune his carburettor to suit his own special requirements.

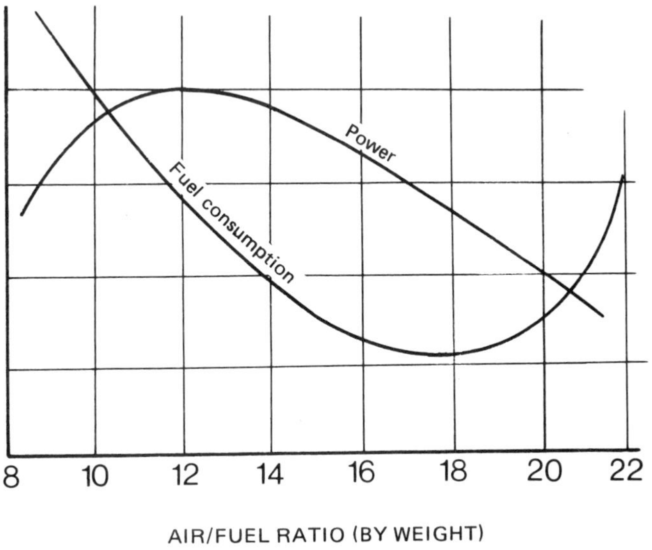

Fig. 2. Curves showing the effect of the variation of the air/fuel ratio on power output and fuel consumption

4 The theoretical carburettor

To summarize, a carburettor is a device which supplies a regulated quantity of liquid fuel and air mixture in correct, automatically controlled proportions.

All modern carburettors make use of a physical phenomenon regarding the flow of fluids. It is the principle of the venturi. Consider the device in Fig. 3.

When a fluid (liquid or gas) flows through this pipe, its velocity (speed) will increase as it approaches, and be at its maximum, at the constriction (throat). This is because the mass flow (weight of fluid flowing past a given point for any particular length of time) must remain constant and the throat presents a smaller cross-sectional area. Since the fluid cannot be compressed it must increase in speed. Now the pressure exerted by the fluid is inversely proportional to its velocity, ie, the pressure decreases as the velocity increases. The minimum pressure (or maximum depression) is therefore at the narrowest part of the pipe.

If a hole is drilled through the wall of the pipe at this point, air would be admitted and would mix with the air flowing through the pipe. If a pipe conveying petrol from a tank is connected to the venturi at this point, the petrol would also be drawn in, as shown in Fig. 4.

Note that the level of the petrol in the tank is just below the outlet from the small pipe into the venturi and that as air flow commences, petrol is drawn upwards due to the suction. The tank is open to the atmosphere. The petrol must be maintained at a constant height relative to the outlet and this is achieved by incorporating a float-controlled inlet valve in the tank. As the tank fills, the float rises, and the valve closes. As the fuel is used and the level drops, the float falls and opens the valve.

If a throttle device is fitted in the main air passage we have a simple carburettor with the means for containing a supply of fuel in a reservoir (float chamber), at a constant level (controlled by the needle valve), for mixing it with the air supply (by the jet), and regulating the supply of mixture (by the throttle valve).

Unfortunately this carburettor has one failing which renders it unsuitable for use on a variable speed and load engine. The orifice (jet) size is carefully selected and calibrated to supply the correct amount of petrol at one particular airspeed, through the air passage. As this airflow is increased by opening the throttle, the airspeed increases and the depression at the jet increases. As gaseous fluids (air) are compressible this reduction in pressure is accompanied by a reduction in density, therefore the weight of air is proportionally less. Liquid fluids (eg petrol), however, are not compressible and, within limits, respond directly to changes in pressure with changes in flow. The net effect is for the mixture ratio to become richer as the airflow is increased, and weaker as it is decreased.

This basic inability to maintain a given mixture ratio over a range of airflows is an inherent feature of the simple, static carburettor and renders it useless for automative applications.

Its construction has been shown in order to illustrate the theoretical laws of fluid flow and their application to a practical carburettor.

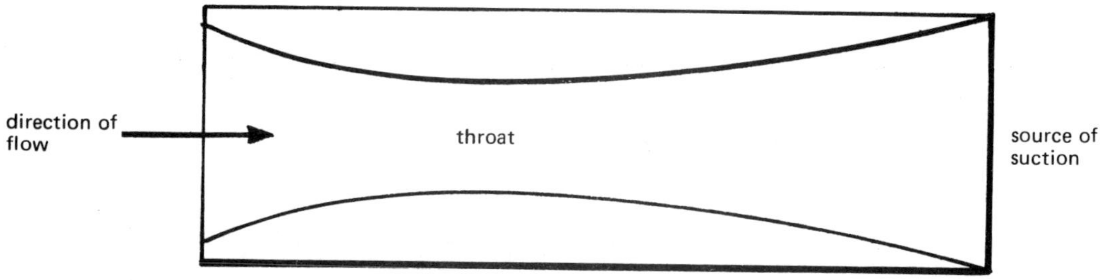

Fig. 3. The principle of the venturi

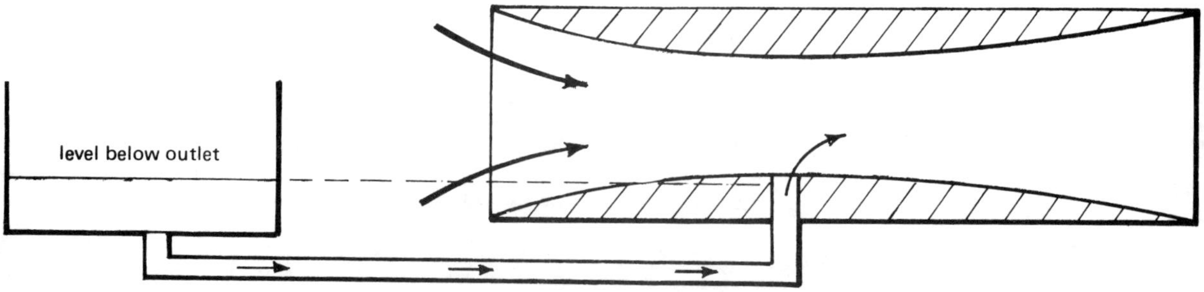

Fig. 4. The principle of fuel flow into the venturi from a tank

Chapter 2 Fundamentals - practical: The SU Carburettor

Contents

1 The Solution – the basic SU carburettor

The basic problem can be overcome in two ways. Firstly, the simple carburettor can be made less simple and more useful by incorporating multiple jets, bleeds and orifices, which, by careful design, can provide a constant mixture ratio over a tolerably wide range of airflows. The instrument is then termed a compensated static, or variable vacuum, carburettor.

The second approach involves a re-appraisal of the fundamental principles, and the introduction of a new concept.

If the airflow can be increased *without increasing the depression* and the fuel flow can be controlled by automatically varying the size of the jet, then the mixture ratio can be maintained constant over a large range of airflows.

The practical expression of this concept takes the form shown in Fig. 5.

The piston presents a constriction in the pipe and forms a 'venturi'. It slides within the chamber which is in communication with the downstream side of the piston, through a port in the piston. Airflow past the piston creates a reduced pressure in the mixing chamber (the part of the pipe downstream of the mixing chamber) which is sensed in the suction chamber. The piston rises to a height which is governed by the magnitude of the depression in the suction chamber balanced by the weight of the piston. These two values must always be in balance. As the piston is free to rise or fall it will hunt for a position where the two forces are in equilibrium. If the airflow is increased, (by opening the throttle), the suction developed in the mixing chamber will be sensed in the suction chamber. As this suction force is greater than the weight of the piston, the piston will rise, increasing the area of the choke. This allows more air to pass but as the choke area is larger *the velocity remains the same* and *the depression is unchanged.* Movement of the piston causes the tapered needle attached to it to be withdrawn from the jet, thus increasing the annular area around it and permitting more fuel to pass. The rate of increase of fuel flow with needle lift is easily regulated by the design of the taper of the needle.

This is the principle of operation of the SU carburettor and it is termed a constant depression, or variable choke, instrument.

For practical operation in various phases of engine running, see Part 2, Type H, Operation.

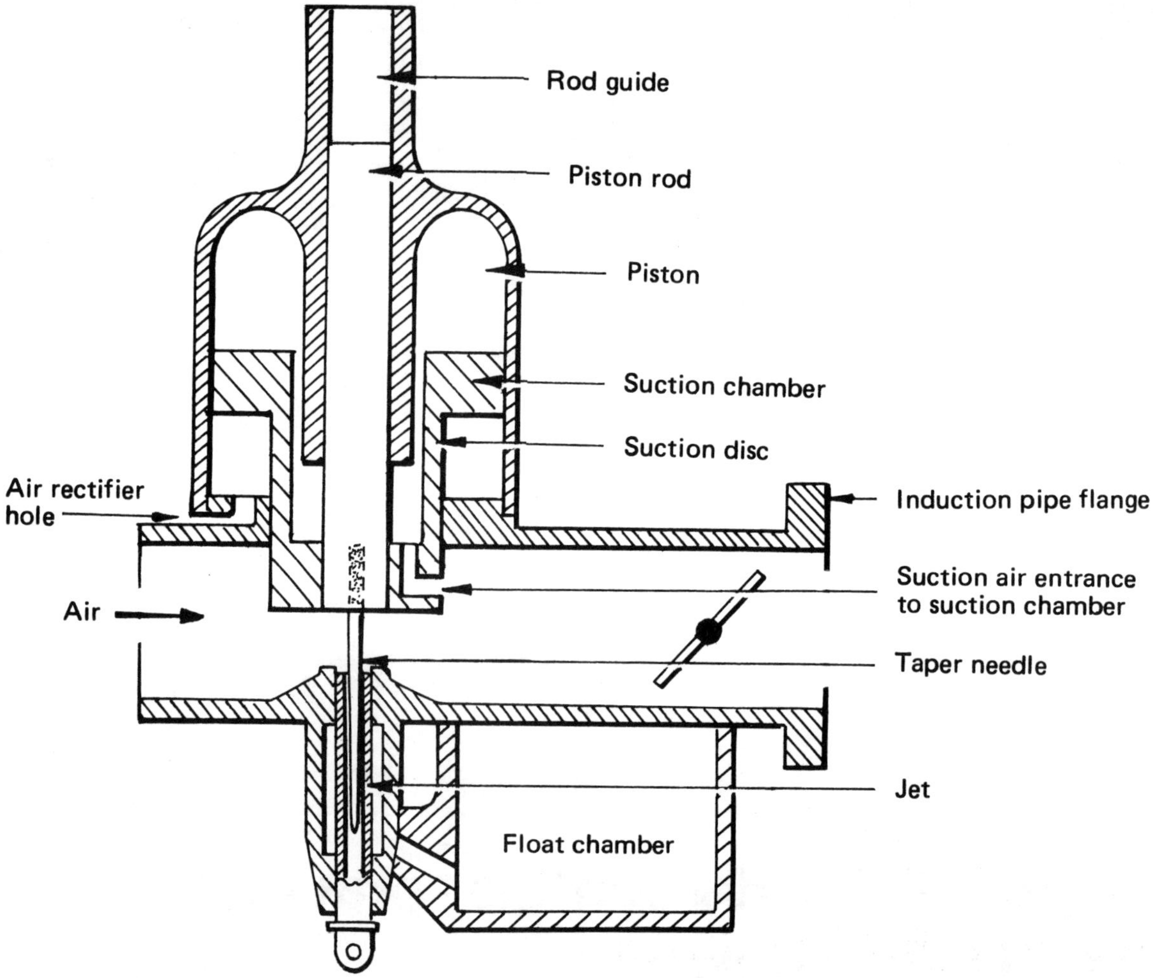

Fig. 5. The basic SU carburettor in simplified cross-section

Chapter 3 Installation

Contents

1 Installation – horizontal carburettors

The four types of SU carburettor dealt with in this manual – H, HD, HS, HIF – are all of horizontal type, ie the bore of the gasway is horizontal, and the axis of the piston and suction chamber is therefore vertical, when mounted on the engine.

Some vehicles which have very restricted engine bay space have the carburettor mounted at an angle, when viewed from the side of the unit, to facilitate installation and/or servicing. The angle, normally 20° or 30°, is achieved by the use of an adapter or curved intake manifold, the carburettor being basically a horizontal type. (The flange faces are still at right angles to the bore). This type of installation geometry is termed 'semi-downdraught'.

Carburettors may be right- or left-handed according to the position of the choke and throttle interconnection. The unit is right-handed if the linkage is on the right-hand side when looking into the inlet, and vice versa. Note that on types H, HD and HS the float chamber may be on either side, and that its position is not relevant in the handing of the unit.

The float chamber is normally mounted in front of the carburettor. In this position flooding at the jet when going downhill, and starving the jet when going uphill, is prevented.

2 Identification of carburettors on multiple installations

To avoid confusion in identifying carburettors on multiple installations, particularly transverse and Vee engine configurations, the following terminology has been established by the manufacturers of SU carburettors and should be used when discussing the location of the unit.

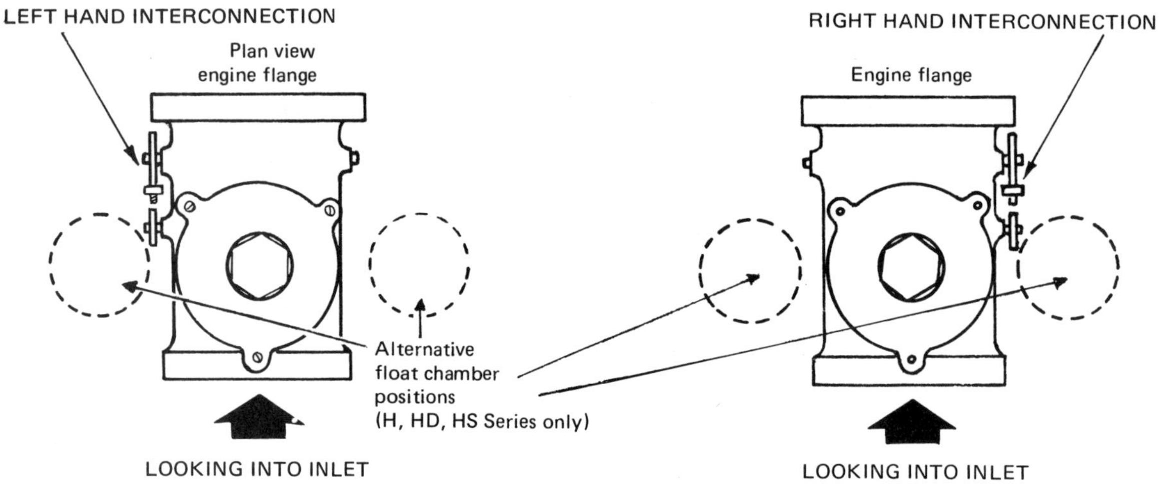

Fig. 6. How SU carburettors are handed

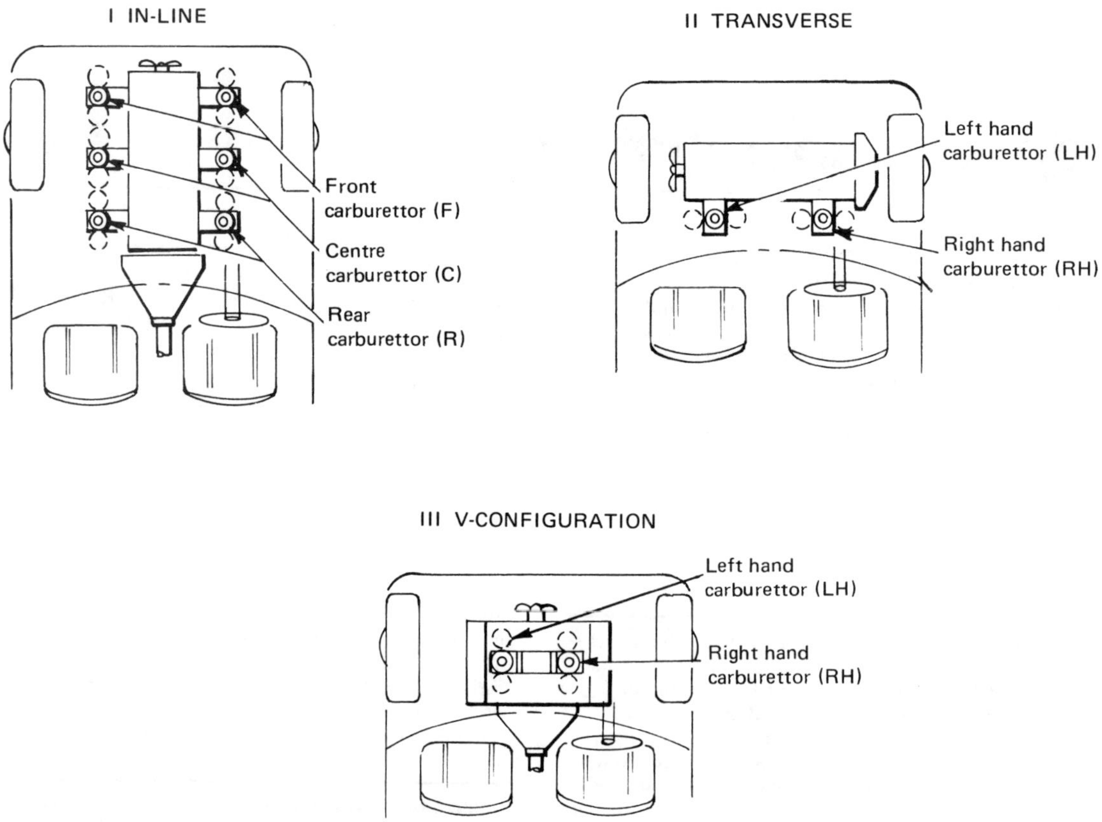

Fig. 7. The manufacturer's description for various types of multiple carburettor installation

Chapter 4 Maintenance and Servicing

Contents

1 Introduction

The maintenance and servicing procedures for SU carburettors of different types are very similar as the units utilize the same basic principles of construction and operation, and embody similar components and features. Maintenance and servicing procedures will therefore be outlined which will apply to any instrument and detailed instructions appertaining to any specific type will be given in the appropriate section of this manual relating to that type. (See Introduction, Part 2, SU Carburettors by Types).

2 Definitions

For the purpose of this manual maintenance is defined as operations necessary to maintain the carburation system as a whole in good repair, without recourse to extensive disassembly of any parts. Because of the simplicity of the work involved it is often overlooked and the next higher level of work, servicing, has to be performed prematurely.

Servicing is defined as operations necessary at regular intervals of time or mileage to check the performance of the carburettor and to restore to specification, if necessary, by adjustment of the facilities provided. It also entails carrying out periodic duties such as lubrication. Failure to observe servicing instructions results in the next higher level of work, overhaul, having to be performed prematurely. In addition, the performance and condition of the engine will deteriorate.

3 Maintenance

Maintenance is limited to visual inspection and should be carried out quite frequently to prevent troubles developing unnoticed. The maintenance period quoted in the car manufacturer's handbook should be adhered to. If no recommendations are given, 3 months or 3000 miles, whichever is the sooner, is quite a reasonable interval for cars operating in normal conditions. Hard driving, or abnormal adverse conditions, such as operation in hot, dusty climates, dictate more frequent intervals.

4 Procedures

First the engine should be thoroughly washed down by brushing on a de-greasing fluid such as Gunk or Jizer and washing off with clean water. Take care to cover any apertures such as oil filler, dipstick hole, air inlets etc to prevent the entry of water. Ensure that no water penetrates into the distributor or coil by covering them with polythene bags. Clean these electrical items, including plug leads, separately, with a petrol soaked cloth.

Now inspect the whole of the induction system from air intake to cylinder head, as follows.

Check that the air intake is not obstructed. If the air supply is taken from a duct carrying air to the car's heating and ventilating system, remove the flexible hose and check that neither it nor its supply pipe is partially blocked with dead leaves etc.

Check flexible hoses for cracking, chafing and punctures. Check that they do not rub against sharp objects or rest on hot surfaces. Check that connections at each end are secure. If screw-drive (Jubilee) clips are fitted, lightly lubricate with thin oil and make a note to replace them at the next service opportunity if the screwdriver slots are damaged.

If open intakes (trumpets or ram pipes) are fitted, check the inside surface for accumulation of dirt and clean if necessary.

4 Check that all hoses are not crimped or cracked, be they rigid or flexible

5 Air filters and silencers

Check that there are no leaks at intake or outlet connections.

Check that the two halves of the metal casing fit together properly and that the air filter is secure on its mountings and is externally undamaged.

5 Check the condition of the air filter; its element and fixings

6 Connection to carburettor

Check that there are no leaks and that nuts are tightened.

7 Inlet manifold

Check that there are no leaks at carburettor flange or engine flange connections. Leaks here often show up as stains on the adjacent metallic surfaces.

If the manifold has vacuum tappings for windscreen wipers, windscreen washers or servo-operated brakes, check the union nuts for tightness. If the connection is a rubber sleeve ensure that it is firmly in place and is not cracked or perished. This applies also to carburettors fitted with vacuum connections. Check the manifold to cylinder head nuts for tightness.

If multiple carburettors are fitted, check the balance pipe on the inlet manifold for security and leaks.

8 Carburettor controls

Check all controls for evidence of corrosion and for freedom of operation. If cable operated check for kinks, fraying or breaking of strands and for integrity of outer sheath.

If an articulated linkage check all nuts for tightness.

Check ball joints for freedom of movement and freedom from excessive backlash or lost motion. Adjust if necessary but do not disturb carburettor settings. Check that all split pins, locking wire or other locking devices are secure.

6 Test the fixing of the carburettor to the inlet manifold; it should be tight but not too tight

9 Carburettor body

Check security of all fixings. Check that nuts and screws associated with controls are all in place.

Check that springs are not fouled or broken.

Check that damper cap is tight.

Check that fuel connection and vacuum connections for ignition control (leading to distributor) are tight.

Inspect underside of carburettor for fuel leaks.

10 Tuning

Tuning is probably the most talked about and least understood aspect of car maintenance.

Carburettors are often spoken of as 'going off-tune'. This is really incorrect, for as we have seen, the carburettor consists of relatively few moving parts which are subject to very low loads and consequently do not wear to an extent which would impair performance until a high mileage has been covered.

What in fact happens is that the engine's mechanical condition deteriorates or its operating conditions alter and the carburettor, being a very sensitive instrument, is adversely affected. As we have seen in Chapter 1, even a change of petrol could influence a carburettor's performance due to an incorrect A/F (mixture) ratio being delivered. Also, extremes of ambient temperature or a change of altitude will influence the running due to a change in air density.

The carburettor is rarely responsible for unsatisfactory engine operation. It does not, of its own accord, 'go off-tune' and poor performance, if it can be attributable to the carburettor, is usually due to poor maintenance, incorrect adjustment, or incorrect assembly after overhaul. It is subject to fuel supply blockage but this is most unusual due to the 'self-cleaning' nature of the variable jet.

Tuning then, is restricted to routine servicing intervals and re-adjustment after any work on the engine or any other change of operating parameters which call for a change in mixture ratio and slow-running (idling or tick-over) speed.

'Fiddling' with controls and adjusting screws in an effort to improve performance, particularly in the case of multiple carburettors, is to be discouraged. It invariably gives poor results and as previous settings have been disturbed and lost, the tuner will have wasted his time and will have to start from the beginning again.

11 Engine condition

It is important that the engine condition is known. Good results will only be obtained if the engine condition is generally sound. If the engine has covered a very high mileage and is well worn, no amount of careful tuning will restore its performance.

A slightly worn but serviceable engine can respond well to carburettor tuning as the adjustments provided can, to a certain extent, compensate for defects. For example, worn valve guides will admit extra air on the induction stroke and dilute the incoming mixture downstream of the carburettor.

This can be corrected by enriching the mixture to compensate for the additional air, but this will only be correct for one engine speed, and at other speeds and/or loads the resultant effects are quite unknown.

Leaking valve seats or piston rings will upset the vacuum at the carburettor giving irregular 'pull' at the jet.

The parts of the engine responsible for 'breathing' are of course the most important from the point of view of carburation. These parts are the cylinder head itself, the valves, valve guides, valve operating mechanism (camshaft, cam followers or tappets, pushrods, rockers, rocker shaft), pistons and piston rings, cylinder bores, intake and exhaust systems. Any known defect

7 Check the vacuum advance connections between the carburettor and distributor

8 All carburettor controls should be well connected and functioning through their full range of movement

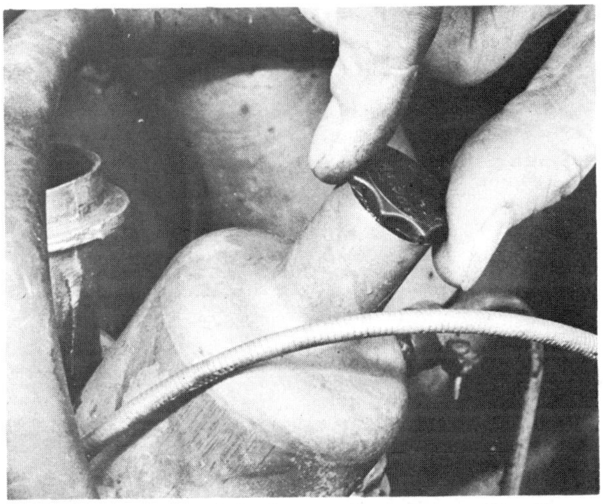

9 Check the security of the damper; also make sure it has the correct oil in

in these components must be rectified before any attempt is made to tune the carburettor.

If the engine is in fairly sound condition it is still advisable to first adjust the ignition timing and set the valve clearances according to the instructions given in the car manufacturer's handbook, and to the data marked on the engine.

The condition of the spark plugs should be ascertained and if necessary they should be replaced with new ones, properly gapped.

A quick, practical method of assessing the condition of the induction system is to carry out a cylinder compression test, using a proprietary compression tester. This instrument is basically a pressure gauge incorporating a non-return valve, which registers and records the maximum pressure generated in the cylinder.

This pressure is termed the compression pressure and is related to the compression ratio which you will need to know. It will be as quoted in the car manufacturer's handbook unless any modifications have been made to the engine such as increasing the ratio to increase performance, or decreasing it to enable a fuel of lower octane rating to be used.

If the compression ratio has been increased it is not possible to determine the exact ratio without stripping the engine.

If it has been decreased this will usually be evident by the presence of a thin metal plate, fitted between the cylinder block and cylinder head, the edge of which may be seen (but do not be confused by the cylinder head gasket). In this case the compression ratio is probably the value given in the handbook for the low compression (LC) model.

Following the instructions given by the manufacturer of the compression tester, take readings from each cylinder in turn and compare the results with the correct compression pressure, and with each of the others.

If the reading from any cylinder is greater than 20% below the correct compression pressure, or if the difference between the highest and lowest readings (there are sure to be differences) exceeds 10% of the correct compression pressure, then the engine should first be overhauled.

From the above it will be seen that balance, or consistency of breathing ability among the cylinders, is as important as a good overall condition.

If the engine satisfies this test, and the instructions given under Maintenance have been attended to, the carburettor is then ready to be tuned.

11.1 Spark plug condition can be used to ascertain the state of tune

11.2 Cylinder compression testing is a quick, practical method of induction system testing

Chapter 5 Overhaul

Contents

1 Introduction

Overhaul is defined as work which is necessary to restore a worn or damaged unit into a serviceable condition by replacement or repair of any or all of its component parts. The unit must be removed, disassembled, cleaned, inspected, repaired as necessary, re-assembled and refitted.

2 Overhaul

Overhaul becomes necessary when the performance of the unit deteriorates and cannot be restored by normal tuning (see Chapter 4, Maintenance and Servicing), or when it has suffered accidental damage due to failure of adjacent parts, fire etc. It is quite distinct from servicing and should not be confused with these routine tasks. The life of a carburettor before overhaul is required, or the time between successive overhauls, is difficult to assess. It depends on many factors of which perhaps the greatest is the attention given to its maintenance and servicing. A unit which is operated with unlubricated spindles, dirty linkages and no air filter will naturally suffer accelerated wear.

The work involved is within the capabilities of the majority of owner-drivers. No extensive workshop equipment is required, most of the work being done with the usual hand tools which should be a part of every private car's tool kit. However, it should not be undertaken lightly, as a complete overhaul takes an appreciable time and at best, entails having the vehicle off the road for a day, if the job is to be done properly. At worst, the delay may be considerably longer, if a defective part is discovered which cannot be repaired and for which a replacement part has to be obtained from the manufacturer or his agent.

Before the work is started the operator should acquaint himself with the detail construction and operation of the unit (see appropriate section of Part 2 of this manual) and familiarize himself with the overhaul instructions. It should also be ensured that all necessary tools, materials and likely spare parts (ie those which are invariably found to require replacement) are to hand. These are itemised next for a typical carburettor. Specific units may require additional items, when these will be called up in the appropriate section dealing with that type.

3 Tools

Spanners, open-ended
Spanners, ring
Screwdrivers – flat blade and Phillips head
Razor blade
Surface plate or thick piece of plate glass
Steel straightedge
Feeler gauges
A large, clean tray
Open-topped containers
Pipe cleaners

4 Materials

De-greasing fluid
Petrol
Clean, lint-free cloth
Emery cloth
Metal polish
Paint (small quantity – for marking)
Thin oil (SAE 20 – sewing machine oil)
Engine oil (SAE 30 or 20W/50)
Tinman's solder and flux
Araldite adhesive and hardener
Hand cleanser
Methylated spirit
Fine grinding paste
Petroleum jelly (Vaseline)

5 Spare parts

A gasket pack
An overhaul pack
Spring washers for mounting studs
Rubber sleeve

Note:
A gasket pack includes all washers and gaskets necessary for a strip, clean and reassembly operation.
An overhaul pack (referred to as a maintenance pack by the manufacturers of SU carburettors) includes jet, needle and seat and throttle disc screws. The pack should be used in conjunction with the appropriate gasket pack to form a basic overhaul kit.
Part numbers for gasket and overhaul packs for any specific carburettor type are given in the appropriate section in Part 2 of this manual.

6 Replacement parts

When ordering any replacement part from the manufacturer or his agent, always give the following information:

a) Make, model and year of manufacture of the car.
b) Engine capacity and number of cylinders

In the case of a multiple carburettor installation identify the unit by reference to its position, as described in Part 1, Chapter 3 Installation.

7 Removal from engine

1 Wash the carburettor and the surrounding area as instructed in Part 1, Chapter 4 Maintenance.
2 Disconnect the air intake system from the carburettor at the intake flange, and remove as much of the intake system as is necessary to give good access to the carburettors.

Warning
If working on a system which incorporates an electrically operated thermostatic carburettor, disconnect both cables from the car battery terminals.

3 Disconnect the throttle control at the butterfly spindle lever end by removing the nut and threaded pin if of the articulated ball and socket type, or by loosening the cable clamping screw if of the cable operated type.
4 Disconnect the choke control by loosening the cable clamping screw.
5 Disconnect vacuum advance pipe, if fitted. This will be a very small bore pipe connecting the distributor to a tapping on the carburettor body. The connection at the carburettor is usually a rubber sleeve which can be pulled off the stub pipe on the body. If the connection is a screwed union, hold the hexagon nearest the body with a spanner to prevent it rotating and unscrew the union nut. Check that the sleeve, or union, as appropriate, is undamaged. A cracked or perished rubber sleeve must be replaced. Blank the open end of the pipe with self-adhesive tape to prevent the entry of dirt.
6 With all controls and ancillaries disconnected, the carburettor is now ready for removal. If two or more carburettors are fitted they should be marked to ensure that they are refitted in their correct positions. It is usually inadvisable to remove any interconnecting linkage, this will come away as the carburettors are detached from the intake manifold, or cylinder head.
7 Remove the nuts and spring washers securing the carburettor to the manifold. Withdraw the carburettor and the thick heat insulating gasket (if fitted), or thin gasket, from the studs. Discard the spring washers having ensured that new replacements are available. Refit the nuts to the studs to prevent their loss. If the unit is to be off the engine for a long period a blanking plate should be made (to the same shape as the carburettor flange, and cut from ·25 in/6.35 mm thick plywood, perspex or aluminium sheet) and fitted to the manifold with the nuts to prevent entry of dirt. Check the nuts for damage, particularly crossed threads and burred corners, and replace if necessary.
8 Inspect the mating faces of the manifold, gasket and carburettor flange. Stains or discolouration indicate that there is leakage due to warped faces. If found, make a note of this; it must be rectified later.

8 Disassembly

Note: When overhauling a multiple carburettor installation, it is advisable to work on one unit only at a time, this will prevent accidental interchanging of parts. Parts should not be interchanged (even if they are identical), as they will have become 'run-in' with their respective mating surfaces. Also, many parts, such as jet and throttle spindle on twin carburettor installations are 'handed'; and some parts may differ between front and rear carburettor.

9 Cleaning

Warning: *Conduct cleaning operations in a well ventilated area and away from naked flame.*

Clean all parts by immersion in a bath of petrol. Keep the small and delicate parts (needle, piston spring, damper rod, piston lifting pin and spring, float etc) separate from the heavier parts to avoid damage. Scrub the body casting inside, and the lid of the float chamber with a toothbrush to clean dirt from the crevices.

Remove all traces of gasket material and cement from the body flange faces with a sharp razor blade.

Allow parts to air dry and place them on a clean, dry tray.

10 Inspection and repair

With all parts cleaned and set out they can now be inspected to ascertain the cause of the unit's poor performance, and to decide which parts are fit for further service, which have to be repaired and which have to be scrapped and replaced by new ones.

Procedures are given for major parts of the carburettors dealt with in this manual (see Part 2).

11 Body of carburettor

1 Check externally and internally for cracks. These are most likely to be found around the bolt holes in the mounting flanges or at the junction of the flanges with the body. If a crack is found the body and, therefore, probably the whole carburettor will have to be replaced, as otherwise the crack will extend into other areas and cause further trouble.

2 Check the flange faces for flatness. A straight edge laid across the face at any angle must contact at all points. If the associated gaskets were marked (see Removal from engine), this indicates in which plane the flange is bent or warped. Flatness may also be checked by laying the flange face on a surface plate or thick sheet of plate glass and checking that it does not rock and that a 0.002 in/0.05 mm feeler gauge cannot be inserted anywhere at the edge between the flange and the plate. If the flange is not flat, true it up by placing a sheet of fine emery cloth on the plate and lightly rubbing the face on it. When an even marking all over the face is achieved, and the checks detailed above are satisfied, the face is as flat as is necessary.

3 Check all tapped (internally threaded) holes for integrity of threads. If any thread is badly damaged the body will either have to be replaced or the thread repaired. The only satisfactory method of repair is the fitting of a thread insert, sometimes called a wire thread insert or Helicoil, according to the proprietary manufacturer. This is an internally and externally threaded sleeve of a much harder material than the original soft aluminium, which screws into the damaged hole after it has been 'cleaned up' by drilling out the remains of the old threads. This repair method is rather sophisticated and requires equipment beyond the resources of most amateurs, but consideration should be given to it in the case of an obsolete carburettor when spares are not available. A good engineering shop should be able to do the job quite cheaply.

There is another method of repair. This is drilling the hole to remove the threads and then recutting an oversize thread. This is effective but requires that the attaching part be non-standard and that the hole through any associated item be drilled oversize to suit.

Most thread damage is caused not in service but by over-tightening screws when assembling. The existence of a stripped thread will at least serve as a reminder later in the operations.

4 Blow through all small holes and passages to ensure that they are unobstructed.

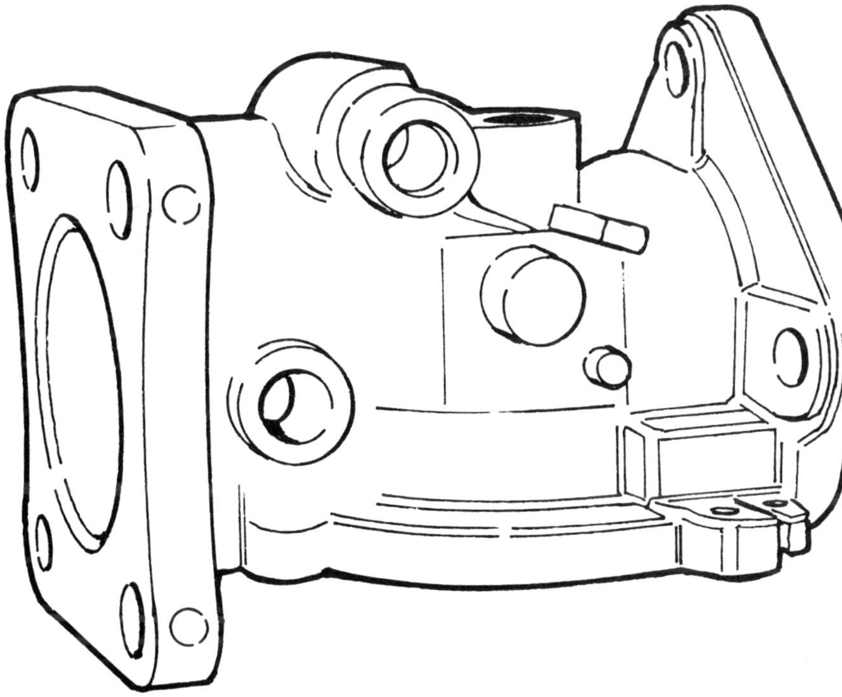

Fig. 8. The basic carburettor body

5 Corrosion is not a common problem with SU carburettors due to the excellent resistance of aluminium and due to their operating in a naturally hot, oily environment, but it is sometimes to be found on a carburettor which has been out of service or removed from the engine for a long time.

Any external corrosion may be removed by light filing or scraping and then finished to a smooth surface with emery cloth.

Internal corrosion may be dealt with similarly but if it is extensive or encroaches upon a critical area (such as the choke bore in which the piston slides) the body will probably have to be scrapped, as excessive clearance here will cause an air leak and upset mixture control.

6 Check the throttle spindle bearings for wear, in conjunction with the spindle. Insert the spindle and *check that it rotates freely but that there is no radial play.* If slackness is perceptible this could be due to wear of the spindle. To ascertain this, slide the spindle out so that the unworn portion of the spindle now runs in the bearing. If the fit is satisfactory the wear must be in the spindle. If the fit is still as loose, the wear must be in the body; and if there is an improvement there is wear in both the body and spindle! This is the feature which most often causes the body to be rejected, as a good leak-free fit here is essential to the proper functioning of the instrument as a whole. The body is only rejected, of course, if the spindle bearings are formed direct in the body, as they are on the smaller carburettors and type HS, and no separate bearings or bushes are fitted. On larger carburettors and types HD and HIF, separate brass or PTFE (plastic) bushes are fitted and these may be easily replaced.

Having isolated the wear, if it is in the body and beyond permissible limits, there are several alternative courses of remedial action.

a) Replace the body with a new item.
b) Replace the brass or PTFE bushes, if fitted.
c) If no bushes are fitted, drill or reamer out the holes, make and fit a pair of brass bushes.
d) Drill or reamer out the holes to an oversize, make and fit corresponding oversize sleeves to the spindle.

Remedy a) is to be preferred where possible. On an obsolete unit, though, for which a replacement part is not available, some repair scheme must be adopted.

Remedy b) is the obvious remedial action if provision has been made for this. The bushes are normally fitted in counterbored holes from the outside of the unit, ie the holes are not bored right through, but a flange is left on the inside to form an abutment and limit the end movement of the bush. The bush is an interference fit in the counterbored hole and is removed by screwing in a tap of suitable diameter to cut a thread and then simply pulling it out with the tap. Clean out the insides of the holes and press in the new bushes which should be a tight fit. They are best fitted by passing a long screw through each bush and the bore of the carburettor body (like an extended, threaded, throttle spindle) and fitting a large washer and nut to each end. Rotation of the nuts will pull the bushes into the bores ensuring parallelism and squareness.

If they are fitted by pressing in with the aid of a vice, or hammering in, there is a risk of damage, or at least of relative misalignment, which may prevent entry of the spindle into both bushes and will accelerate wear.

Finally, a drill or reamer should be passed through both bushes to ensure that the bores are in line, removing metal if necessary.

Remedy d) is not a recommended repair as the concentricity of the spindle is difficult to maintain. It should only be used when the bearings and spindle are very worn and neither part can be obtained as a spare.

The bearing holes in the body must be enlarged to the minimum size necessary to produce a circular hole, concentric to the original hole and of good internal surface finish. The enlargement should be done with a drill or reamer approximately 0.0625 in/1.58 mm larger than the spindle size. Oversize brass sleeves, similar to the bushes in the previous example, are then machined and soldered or bonded (with Araldite) to the original spindle. With careful workmanship a satisfactory job may result.

12 Spindle

Wear of the spindle journal diameters (those which run in the bearings) is the most likely cause for rejection of this item.

If the journals have a polished appearance they are fit for further service.

If there is a visible reduction in diameter, or a perceptible ridge can be felt with the fingernail, it must be replaced or repaired. Two repairs are possible.

Firstly, fitting brass sleeves as described just previously when the body must also be reworked, and secondly, by tinning the spindle with solder and then finishing circular and smooth with a fine file and emery cloth. Silver solder is preferable as it is harder and more wear resistant. Neither of these repair methods should be used if a replacement part is obtainable.

Check the tapped holes and the external threaded portion for condition of threads.

Check the slot for cleanliness.

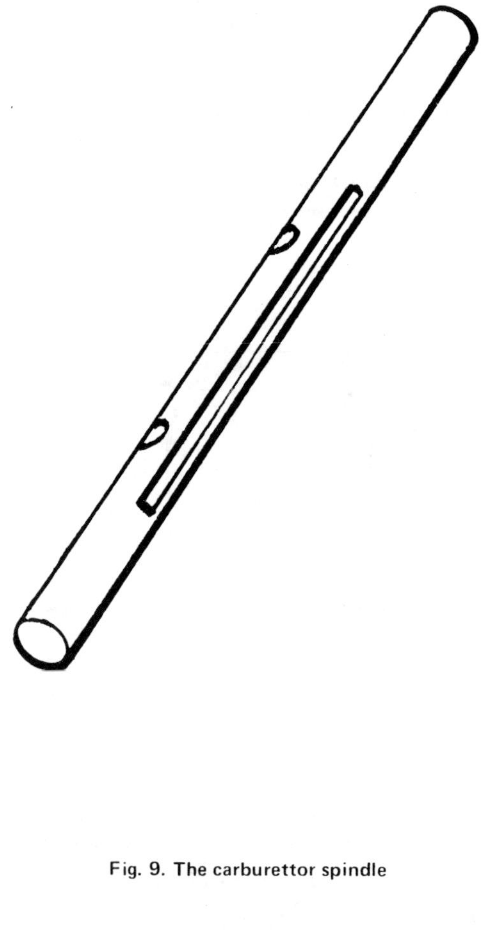

Fig. 9. The carburettor spindle

13 Butterfly plate

The butterfly plate is not subject to any wear and need only be inspected for corrosion and condition of the edge. The edge is not machined at right angles to the surface, nor is the plate itself circular. No attempt should be made to alter the angles or the profile as this will destroy the 'valve closed' sealing. If the edge is nicked or burred due to the plate having closed on some foreign object, the mark must be removed very carefully with a fine file.

No repair to the overrun valve (when fitted) is feasible. A damaged valve seat or spring entails the replacement of the complete valve plate.

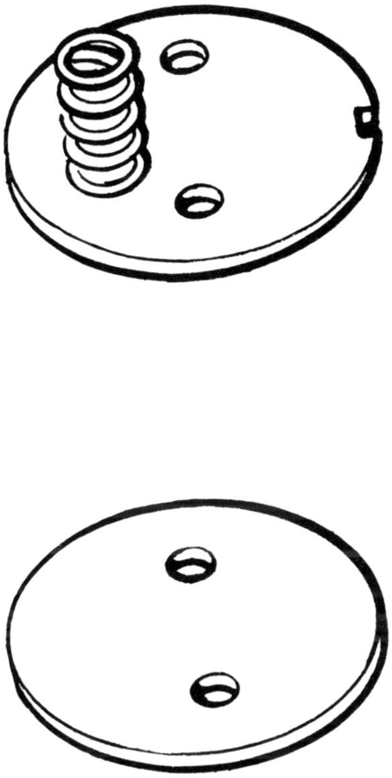

Fig. 10. Two types of butterfly: one has an over-run valve

14 Piston lifting pin and spring

The piston lifting pin is unlikely to be damaged in any way. Check that the spring is not distorted or corroded. Replace if defective.

15 Jet holder, washer, adjusting nut and spring

Check these for mechanical damage and for corrosion. Check the condition of the threads and see that they screw together easily.

16 Needle

Check the needle for wear. As it does not rub against any component in its operation, wear will only occur if it has been assembled eccentrically relative to the jet. Wear will show as linear marks or scratches. Replace the needle if any are visible. *If the needle is replaced due to wear, replace the jet also as this will be worn.*

Check for straightness. Roll the shank (the plain, parallel portion at the top) across a flat surface and watch the path described by the point of the needle. If it rises and falls the needle is bent and must be scrapped. If its height remains constant this proves that the point is concentric with the shank, but does not prove that the needle is straight throughout the remainder of its length. Check the gap between needle and surface plate with feeler gauges at a point midway along its length, rotate the needle through half a turn and check again. If the gap is the same the needle may be considered straight. A bent needle is usually obvious, but if any doubt exists it should be scrapped and replaced with a new one. Do not attempt to straighten the needle.

Check that the shank is in good condition and that it is a good fit in the piston with no trace of sideways movement.

A letter code is stamped on the shank. Check that this corresponds with the recommended needle for the particular application (refer to Appendix 1).

The needle should only have been changed as a result of extensive engine modifications or for operation at elevated altitudes (see also Notes on Needles, Appendix 2).

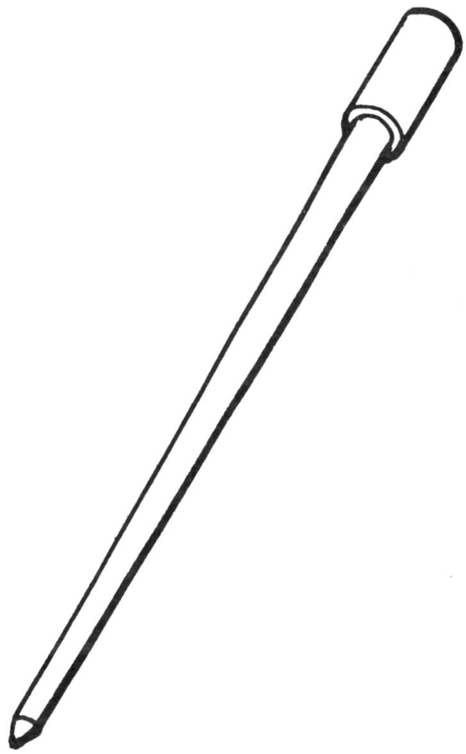

Fig. 11. The needle

17 Jet assembly

1 *Type H*

Check that the jet is a good sliding fit in the jet holder (glands removed from jet holder).

Check the steel fork end for corrosion and clean up if necessary with a fine file. Take care not to erase the jet identification number which is stamped on the face.

Check the linkage pin hole for wear and fit the pin and associated linkage to test for freedom of movement. If the hole is worn it can be drilled oversize and an oversize pin fitted.

Blow through the jet bore to ensure that it is unobstructed.

2 *Type HD*

Check that the jet is a good sliding fit in the jet holder.

Clean the diaphragm very carefully with methylated spirits and inspect closely for the following defects:

Cuts
Cracks (these may be revealed by folding the diaphragm)
Punctures
Bulges
Flattening of the circumferential rib
Tears around the four bolt holes
Distortion

If any of these are present, the jet assembly must be scrapped and replaced. The diaphragm is secured to the jet by clamping between the flange and the collar underneath. It cannot be removed or fitted in isolation and is therefore not available as a spare. Check that the jet cannot be rotated within the diaphragm. If this is possible the jet assembly must be replaced.

Blow through the jet bore to ensure that it is unobstructed.

3 *Type HS*

Check the condition of the threads of the brass union nut and see that the hexagon has not been rounded by the use of an incorrect size spanner, or other tool.

Withdraw the spring sheath from the feed pipe and check that the coloured plastic identification sleeve (or sleeves) is still in place. (For notes on identification of jets see the appropriate section of Part 2, SU carburettors by types). The spring forms a protective sheath for the relatively vulnerable plastic feed pipe, and supports it.

The jet tube assembly consists of an accurately machined brass tube (the jet), a flexible feed pipe and a plastic moulding into which both items are inserted. They are secured by contraction of the moulding by means of metal collars. Short, brass reinforcing sleeves are fitted into the ends of the flexible pipe to prevent it being crushed. The assembly is factory built and may be considered as an integral item; no component part may be individually replaced, and consequently the parts are unobtainable as spares.

The most common fault is looseness of either the jet or the pipe in the moulding. A temporary cure may be effected with Araldite, or other high strength adhesive, but the only practical action is the replacement of the assembly with a new one. Petrol leaks in this area (often above the hot exhaust manifold) are dangerous, and no chances should be taken with suspected defective items.

The plastic moulding and tube are both subject to age-hardening. The moulding becomes brittle and vulnerable to cracking, and the tube becomes hard and loses its flexibility. Inspect the moulding for cracks in the boss on the side, into which the self-tapping screw is fitted to secure the choke (jet lowering) lever. This is particularly susceptible to damage if an oversize screw has been used which tends to expand the boss. The moulding may also have suffered accidental damage as a result of mis-handling of tools in the vicinity of the carburettor.

Note: The foregoing comment regarding age-hardening of plastic items is in no way a criticism of the design. Age-hardening is a physical phenomenon and is an inherent characteristic of these materials.

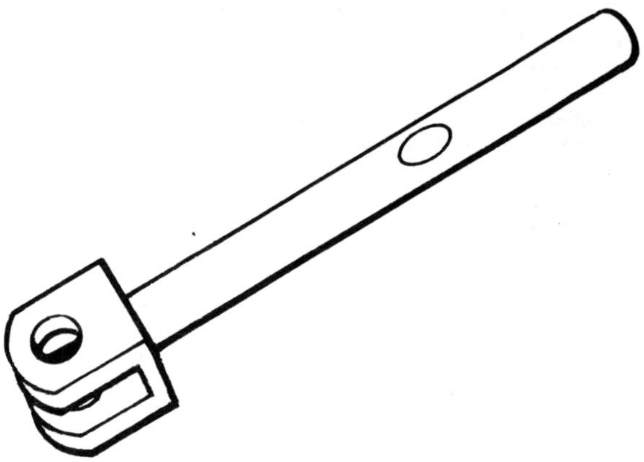

Fig. 12. The Type H jet

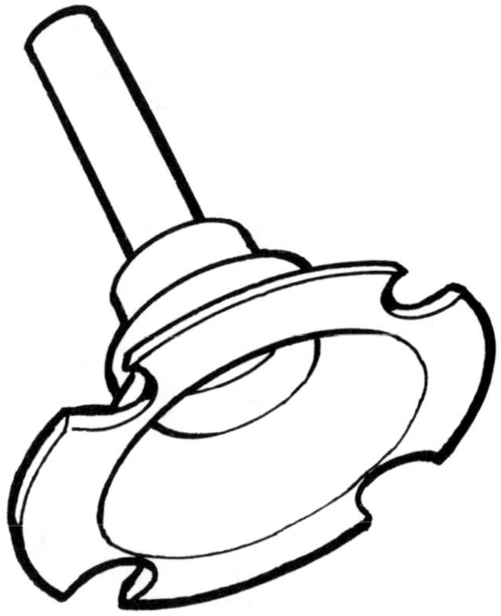

Fig. 12a. The Type HD jet

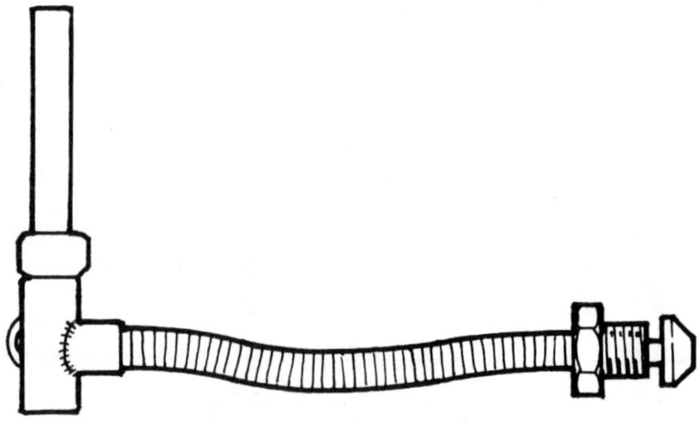

Fig. 12b. The Type HS jet

Check the jet for straightness by rolling it along a flat surface. A bent jet must be replaced; do not attempt to straighten it. Wear marks on the outside of the jet are of little importance provided the jet is a good sliding fit in the jet holder.

Blow through the jet bore to ensure that it and the feed pipe are unobstructed.

18 Piston and suction chamber

Before inspecting the piston and suction chamber ensure that they are scrupulously clean and dry.

Check the lower diameter (the part which forms the valve in the bore of the body) of the piston for scores. Do not attempt to remove these scores – this would involve the removal of too much metal and cause an air leak – but remove any burrs formed at the sides of such marks. This damage is caused by grit becoming embedded in the soft metal of the body, usually as a result of the engine being run with the air filter removed or damaged. Check the associated air intake system, if this type of damage is extensive.

Check the condition of the threads in the tapped hole for the needle locking screw.

Inspect the inside of the suction chamber for corrosion and scoring.

Examine the suction chamber for cracks. These will most likely be found at the roots of the fixing lugs at the base, and entail replacement of the item.

Check around the circumference of the base and on the lugs for file marks made at a previous overhaul to indicate relative position of the chamber to the body. Remove them by careful filing and check that they have not caused cracks.

Check that the lugs are not bent and that the base of the chamber is flat, by placing it on a surface plate and checking that there is no gap at any point on the circumference. If the base is so damaged, restore flatness by lightly rubbing it on a sheet of very fine emery cloth laid on the surface plate. Remove only the minimum of metal necessary to obtain a good surface. Remove the burrs which will be created at the edges. Relieve the inside edge to a very slight chamfer, to assist assembly to the body.

Some units feature dashpots which are vented to the inside of the suction chamber by a drilling, instead of via a pin-hole in the damper cap. These units are identified by an undrilled cap with no provision for venting the spindle bore, and a rib cast on the outside of the chamber at the base of the spindle bore extension. This rib is drilled with a very fine hole which is easily obstructed. To check that this is clear, place a finger over the end of the spindle bore (inside the chamber) and blow through the top.

Check the condition of the threads at the top of the spindle bore extension and clean up, if necessary, with a sharp, pointed instrument such as a scriber. These threads should not be damaged as there is no need for the damper cap to be more than finger-tight.

Ascertain the correct relative positions of the piston and suction chamber from the marks made on disassembly. Lightly lubricate the piston rod with thin oil (household or sewing machine oil) and insert it into the bearing in the suction chamber without rotating it (it does not rotate in service), until the piston disc is about to enter the chamber. Check that the piston rod slides freely in the bore. With the piston in this position check the squareness of the chamber base to the bore by holding up to the light and noting whether the gap is uniform all round the base. This check is particularly important if the suction chamber base has been re-worked, as above.

Push the piston further into the chamber. The piston disc may be felt or heard to strike the edge of the chamber as it enters. This indicates wear of the piston rod and bearing. It is not however, cause for rejection, as at this position the piston is not yet at its lowest working position. Push the piston further in until the bottom of the disc is approximately 0.125 in/3 mm inside the bottom of the chamber. This is the lowest working position. From this position, move the piston through the full extent of its travel without rotating it.

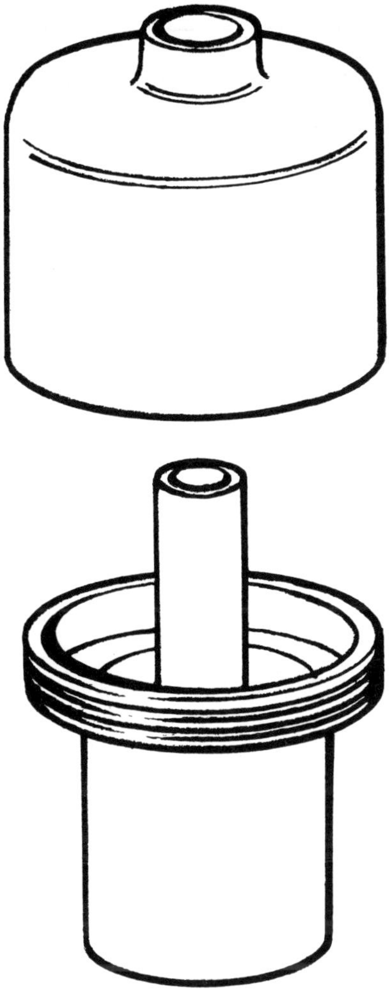

Fig. 13. The suction chamber and piston

If the movement feels 'gritty' and a ringing noise is heard, this is due to dirt between the piston disc (probably trapped in the grooves between the lands) and the chamber. Remove the piston, clean both parts and test again. If the rod and bearing are very worn the piston disc may be touching the chamber. This is incorrect and must be rectified.

It is not always appreciated that the piston disc does not slide in contact with the chamber. The assembly is designed to function with a very small annular gap between the two components. This gap is established during manufacture by accurate machining of the two parts. As a result of wear of the piston rod and bearing the piston is permitted to move laterally and the disc touches the chamber. This causes friction during operation which interferes with the movement of the piston.

Ascertain where the piston is touching the chamber wall by marking the inside of the chamber with a soft pencil and very carefully ease the area by polishing with metal polish. Take care to remove only the minimum amount of metal to give a free (clearance) fit or the designed leakage area will be greatly increased and the efficiency of the unit will be impaired.

19 Spring

The piston spring is lightly loaded and its performance, in terms of elasticity or rate, does not seem to deteriorate with use. The most likely defect is mechanical damage – bowing of the spring or flattening of individual coils – due to careless handling. Check for this visually by rolling it across a flat surface.

Check also for corrosion. This is usually only apparent on a unit which has been out of service for a long time.

The spring is colour coded to identify its size and rate. If the paint colour on the end coil is still visible, check that it is correct for the particular application (refer to Appendix 1).

If it is desired to check the performance of the spring, this may be done by applying a reference load and measuring the overall length, according to the following data.

Do not attempt any repair to a spring. Replace if damaged.

Paint colour on end coil	Load at (oz)	Length (in)	Part Number
Black and green	*5¼*	*2.500*	*AUC 5028*
Light blue	*2½*	*2.625*	*AUC 4587*
Red	*4½*	*2.635*	*AUC 4387*
Yellow	*8*	*2.750*	*AUC 1167*
Green	*12*	*3.000*	*AUC 1170*
Brown	*14*	*3.000*	*AUC 1168*
White	*18*	*3.562*	*AUC 1166*
Red	*4½*	*1.530*	*AUD 4355*
Yellow	*8*	*1.530*	*AUD 4398*
Red and yellow	*24*	*4.812*	*AUC 4478*
Red and white	*40½*	*3.875*	*AUC 4869*
Red and green	*11¼*	*3.875*	*AUC 4826*
Light blue and black	*4½*	*3.875*	*AUC 2107*
Light blue and red	*18*	*3.875*	*AUC 4818*

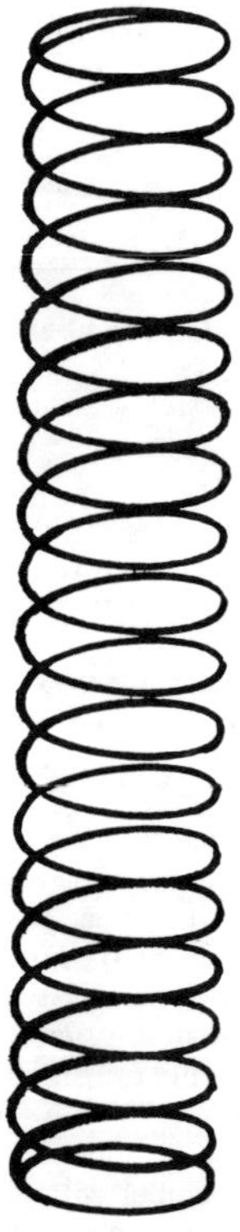

Fig. 14. The piston spring

20 Damper assembly

Visually check that the damper rod is straight by rolling it across a flat surface. Check that the damper cap is square to the rod. The cap, when screwed home, determines the position of the rod and if it is not square to the rod, the bottom of the damper will foul the inside of the piston rod.

Check that the damper cap has a vent hole, and that it is unobstructed, if the suction chamber is not vented (see Piston and suction chamber). If the suction chamber is vented by an internal drilling, the damper cap should be undrilled.

Check condition of the damper cap threads, and of the soft washer under the head if it is to be re-used. Ensure that the valve assembly is perfectly clean. No repairs are possible to this component. Replace a damaged part.

Fig. 15. The standard metal headed early type damper

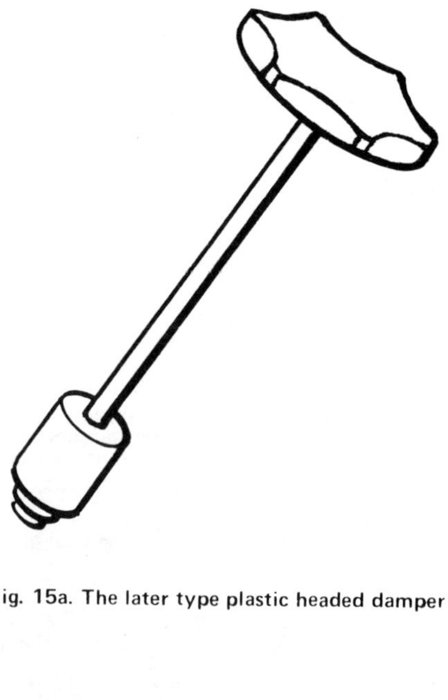

Fig. 15a. The later type plastic headed damper

21 Float chamber

1 Type H

Remove all deposits and corrosion from the inside, and wash out with petrol. Clean out the drilling leading from the bottom of the chamber to the securing bolt hole with a pipe cleaner.

Ensure that the central float guide rod is secure in the chamber base, that it is free from corrosion and that the threads at the top are in sound condition.

Clean up the top and bottom surfaces of the securing bolt boss to ensure a leak-free fit.

2 Type HD

Remove all deposits and corrosion from inside the chamber and the jet spring cup, and wash out with petrol.

Clean out the drilling leading from the bottom of the chamber to the jet spring cup with a pipe cleaner.

Ensure that the plug in the side of the cup is secure. If loose or missing, make up a new plug from aluminium bar, and cement in place with Araldite.

Ensure that the central float guide rod is secure in the chamber base, that it is free from corrosion and that the threads at the top are in sound condition.

Clean all traces of corrosion, deposits, rubber and cement from the top face of the spring cup, with a sharp razor blade. This is the lower sealing face for the diaphragm and must be perfectly clean and flat.

3 Type HS

Remove all deposits and corrosion from inside the chamber. Ensure that the rubber washer has been removed from the inside of the petrol feed pipe hole.

Check for cracks, particularly around cover-attachment screw lugs.

Check all threads. If damaged, refer to Inspection and repair – body, for notes on suggested repair.

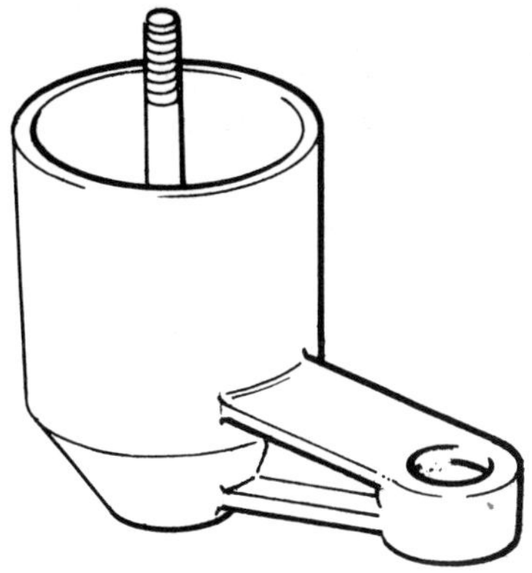

Fig. 16. The type H float chamber

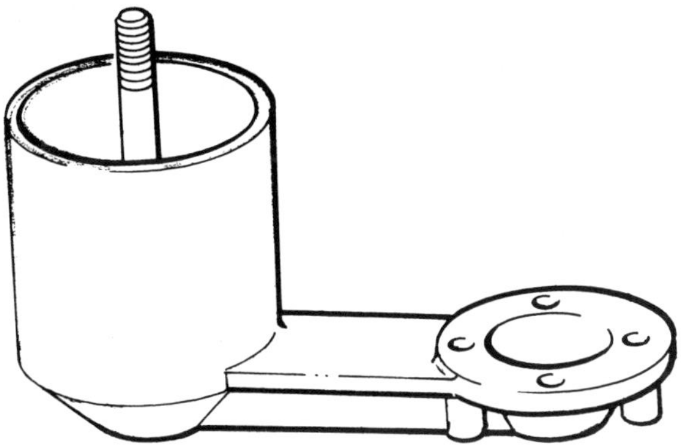

Fig. 16a. The type HD float chamber

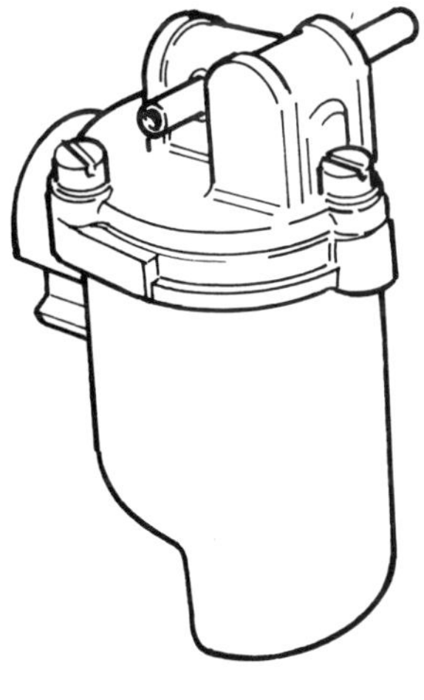

Fig. 16b. The Type HS float chamber

22 Float

1 Type H and HD

Remove all corrosion and deposits.

Check visually for punctures. If any doubt exists, immerse the float in boiling water. A puncture will be revealed by a stream of small bubbles issuing from it. The bubbles are caused by expansion of air or boiling off of any entrapped petrol.

Repair punctures as follows;

a) Thoroughly clean areas around puncture with fine emery cloth.
b) Immerse the float in boiling water until bubbles cease to issue from the hole. Remove float while water is still boiling. (Do not allow water to cool while float is immersed or water will enter the hole).
c) Using a fine soldering iron, apply tinman's solder to seal the hole. Take care to apply the minimum solder necessary to seal the hole, and remove any excess with a very fine file and emery cloth. (Too great an increase in weight of the float due to solder will cause it to sink lower in the float chamber, giving a higher fuel level with possible flooding at the jet).

2 Type HS

Remove all corrosion from the steel lever.

Check that the lever moves freely on the hinge fit and that it is retained securely by the centre tongue. Carefully bend the tongues if necessary to achieve this fit. Check that the lever is securely bonded to the float. Over-enthusiastic bending of the lever to set the correct height may have loosened it. Replace a damaged float assembly. Repairs to this assembly can be effected with Araldite.

Discolouration of the float is quite normal and may be ignored.

Fig. 17. The float

23 Re-assembly

1 Refer to appropriate section in Part 2 of this book.
2 If any rework has been carried out on any component, ensure that it has subsequently been cleaned so that it is free of swarf, metal dust, grinding paste etc.
3 Ensure that the identity tag is refitted. This is very important but is often overlooked, as it appears to perform no useful function. When working on multiple carburettors, ensure that the correct tag is fitted to each unit. If the precaution noted at Disassembly has been observed this will be automatic.
4 It is advisable to fit all new non-metallic parts which are subject to crushing. This includes all gaskets, fibre washers, soft copper washers, Tufnel or composition washers, sealing glands, rubber bushes and spacers etc. These parts acquire a permanent 'set' when in contact with another part for long periods and may not always seal after having been disturbed.

24 Refitting to engine

1 Remove the nuts from the mounting studs in the inlet manifold, or cylinder head. Remove the blanking plate if this has been fitted.
2 Check that the mounting flange face is clean and dry.
3 Fit the correct gasket (thin sealing type or thick sealing and heat insulating type). Do not use jointing compound on either side, it is unnecessary if the mating faces have been dealt with as previously described. Take care not to damage the holes in the gasket when passing it over the threaded portion of the studs.
4 Fit the carburettor, new spring washers and nuts. Tighten the nuts to the torque loading given in the car workshop manual. When this is not quoted, use a normal length ring spanner and *do not use excessive force.* Sufficient to compress the spring washers and a little more is all that is required. This is particularly important when using a thick gasket as it compresses very readily and the carburettor mounting flange is distorted or cracked in consequence.
5 Offer up the inlet flange gasket to the flange and check that the piston chamber vent holes correspond, (it is sometimes possible to fit these inverted, thus blanking off the holes). Fit the gasket and the inlet manifold, ram pipes or silencer with the correct bolts and nuts and new spring washers. Reconnect the remainder of the air intake system.
6 Reconnect the throttle linkage. If of the articulated rod type it may be necessary to adjust the length of the rods by screwing in or out of the socket ends. After adjustment, lock with locknuts. Check that there is no undue load on the linkage when the throttle is closed and that it is not prevented from closing by the linkage. Operate the accelerator pedal to check that the throttle opens fully to the stop and set any stops on the linkage to limit its further movement.
7 Refer to Tuning in appropriate section of this manual.

PART 2
Type H

Contents

1 Introduction

The Type H (Horizontal) carburettor is the earliest of the four types described in this manual. For the purpose of instruction it may be considered as the basic form of SU instrument, since all later types are largely refined versions and all use the same basic elements.

It was most popular in the early 1950s and was fitted as standard to many production saloon and sports cars, and as a 'performance modification' to others. It is often to be found fitted with a thermostatic carburettor on larger cars with a multiple installation.

2 Construction

The unit consists of a die-cast aluminium body in which is formed the choke bore, a suction chamber and piston assembly mounted vertically above it, a jet assembly mounted co-axially with the suction chamber below it, and a float chamber assembly attached by an arm to the body.

The body has a flange at each end, drilled for bolts to mount the unit to the engine and for the attachment of an air cleaner or intake system. A circular, machined platform on the upper side mounts the suction chamber, with a large bore intercepting the choke bore at right angles in which the piston moves.

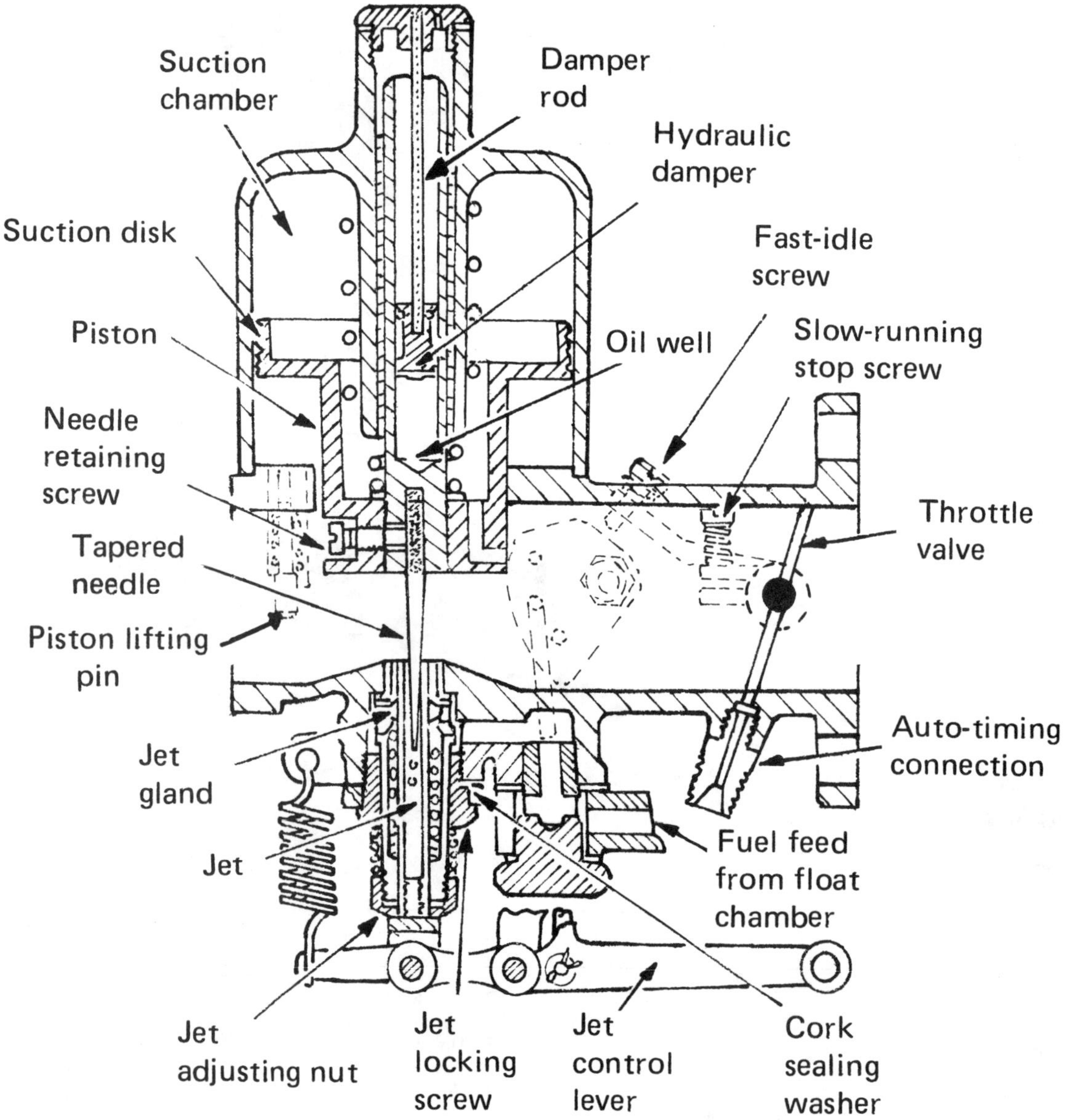

Fig. H2. A cross section through the basic Type H carburettor

The underside of the body has an internally threaded boss for attachment of the jet assembly, and a similar boss for attachment of the float chamber arm. They are in communication with each other by a drilling through the cast web between them.

Other cast features on the body include bosses bored to form throttle spindle bearings, a lug with a hole to accept the tail of a jet return (tension) spring, a housing for a piston lifting pin, tapped holes for vacuum connections, and bosses for throttle/choke interconnecting linkage bearings.

The suction chamber is a die-cast aluminium cylinder with a central guide tube fitted with a hardened steel sleeve bearing. The guide tube is extended at the top of the chamber and internally threaded for attachment of a hydraulic damper. Lugs are provided at the chamber base for attachment to the unit body with screws.

A piston slides in the bearing in the suction chamber, the smaller, lower diameter protruding into the bore of the carburettor body. The piston consists of two parts, the piston rod and the piston disc with integral suction disc. The piston rod is hollow to act as an oil well for the hydraulic damper (which is secured to the suction chamber). The piston has two ports drilled from the top and emerging on the downstream (throttle plate or engine) side of the piston. A tapered needle is fitted to the base of the piston assembly and secured by a locking screw.

3 Jet assembly

The jet assembly consists of the jet (1) which slides in upper bearings (13) and lower bearing (7). Jet glands (12), conical washers and washers (11) are fitted to the bearing and loaded to their respective ends by compression spring (10). The assembly is fitted to the unit body with nut (4) sealed with washer (5) and cork washer (6). The bearing is a loose fit in the nut, thus permitting slight radial movement of the jet assembly to centre it, relative to the needle.

The assembly is held to the body by the clamping action of the nut against the flange at the top of the lower bearing. The soft (copper) washer (5) forms a seal between the lower jet bearing (7) and the body.

The maximum height of the jet is determined by the position of the jet adjusting nut (2), the jet being held in contact with it by the spring loading on the external jet control lever linkage.

The jet size is identified by a code number stamped on one face of the steel fork (see illustration).

When a thermostatic carburettor is used in conjunction with Type H unit/s, the jet is not required to be manually operated and is therefore fixed in a simpler, leak-proof manner. The assembly is referred to as a sealed jet base. Access to the jet, for adjustment, is gained after removal of the cap nut.

The damper consists of a cylindrical, non-return valve assembly retained on the end of a rod by a circlip. The rod is serrated at the top and pressed into a screwed brass cap by which it is secured to the suction chamber. The valve is immersed in the oil in the hollow piston rod and is so arranged that oil can pass freely through it in a downwards direction (piston falling) but presents a high resistance to upwards flow (piston rising).

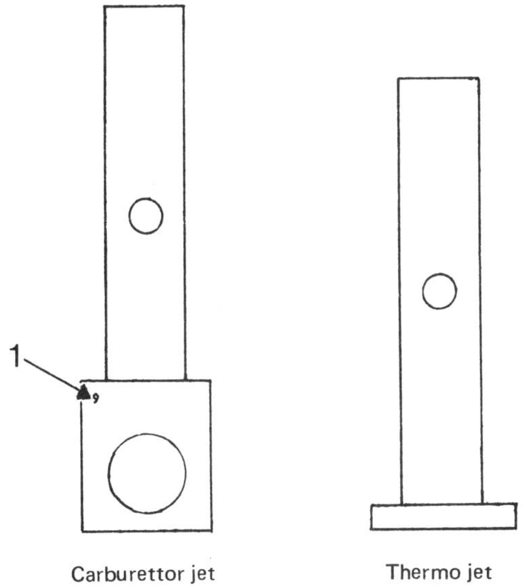

Fig. H3. The two type of jet - standard and thermo

1 indicates the jet size stamped on the jet.
Jet size 9 = 0.09 in 1 - 0.1 in, 125 = 0.125 in

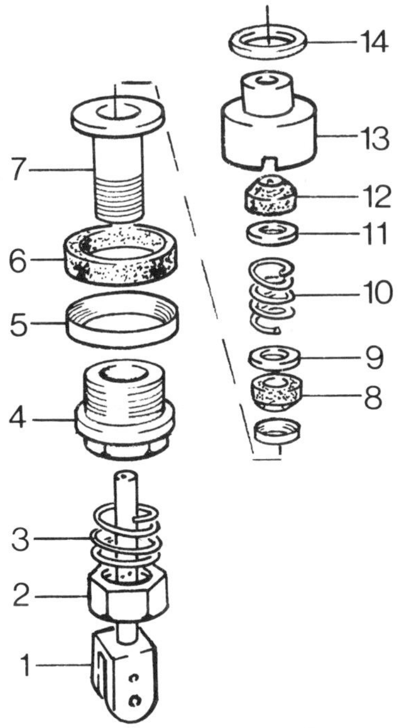

Fig. H3a. The complete jet

1	*Jet*	*8*	*Jet gland*
2	*Jet adjusting nut*	*9*	*Washer*
3	*Spring*	*10*	*Spring*
4	*Fitting nut*	*11*	*Washer*
5	*Copper washer*	*12*	*Jet gland*
6	*Cork washer*	*13*	*Upper bearing*
7	*Lower jet bearing*	*14*	*Washer*

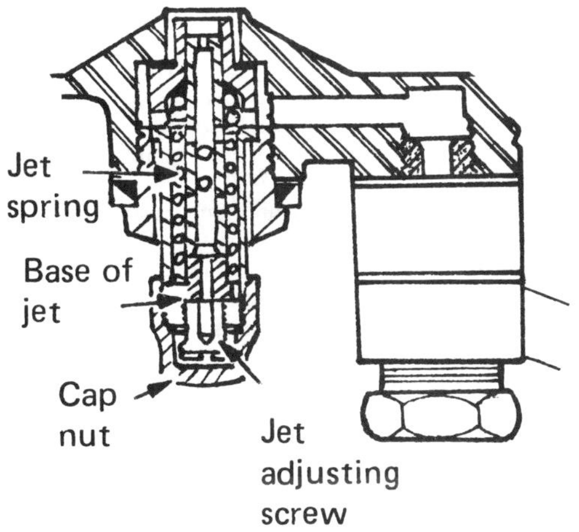

Fig. H4. Sealed jet base with the thermostatic carburettor

4 Float chamber and its operation

The float chamber assembly consists of the chamber, a detachable lid incorporating the inlet valve, and a float.

The chamber is a die-cast aluminium cylinder fitted with a central float guide rod. An extension arm at the bottom of the chamber is enlarged at its outer end and drilled for a banjo bolt which secures it to the carburettor body. Fuel is supplied to the carburettor through a drilling in the arm.

The detachable lid incorporates lugs which carry the float lever hinge pin. The inlet valve consists of a brass seating screwed into the lid, in which slides a steel needle. The position of the needle is controlled by the float lever movement. Fuel inlet to the valve is from an external banjo union secured with a banjo bolt to a tapped hole in the lid. A thimble shaped filter is retained in the inlet by the banjo bolt, the open inlet end being loaded towards it (to seal the periphery) by a conical spring surrounding the filter. The inlet assembly is sealed with soft fibre washers.

The top, threaded portion of the float guide rod projects through a central hole in the lid and retains it with a cap nut. A cover cap is fitted under the cap nut to protect the annular vent from the ingress of dirt. The lid is located in the chamber by a machined spigot at the rim.

The float consists of two brass spinnings soldered together with a brass tube running through the centre. The float slides vertically on the guide rod fixed to the float chamber base and operates the inlet valve via the hinged lever.

Fuel supplied by an external mechanical or electric pump enters the float chamber inlet and passes through the fine mesh filter. As the fuel level in the chamber rises, the float rises and operates the inlet valve, reducing the flow and stopping it when a previously determined level is attained. Petrol flows from the base of the chamber through the drilling in the arm to the jet assembly and rises to the same level (just below the top of the jet when in its fully raised position).

5 Cold starting

To start the engine from cold the external, facia-mounted, mixture control (choke) is pulled out, which lowers the jet assembly in its bearings and glands. With the piston seated on the bridge of the carburettor this lowering of the jet increases the annular area between the needle and the orifice at the top of the jet. The throttle plate is partly opened by the mechanical interconnection between the mixture control linkage and the throttle lever.

See Part 1, Chapter 1 for explanation of the constant-depression concept.

With the engine turning over by hand cranking or by the starter motor, the depression generated in the inlet manifold is sensed in the suction chamber via the passages in the base of the piston. The resultant upward load on the piston assembly, due to the differential pressures sensed above and below the suction disc, lifts the piston against the spring load.

The depression created by the flow of air between the underside of the piston and bridge lifts fuel from the jet. The fuel is atomised in the mixing chamber (that part of the choke bore between the piston and the throttle plate) and passes to the engine, when it should start and run.

When the engine is running the car may be driven straight away.

6 Normal running, full throttle and cruise conditions

With a cold engine the mixture must be enriched to compensate for the effects of fuel condensation on the cold surfaces of the induction system, which has the effect of weakening the effective mixture delivered to the engine combustion chambers. This is achieved by lowering the jet assembly, thus increasing the annular area around the needle at any position of the needle, and permitting more fuel to flow.

As the engine warms through, the mixture strength is weakened by progressively raising the jet assembly until, when normal working temperature is attained, the jet is at its highest position, dictated by the position of the jet adjusting nut which will have been pre-set (refer to Tuning).

At cruise condition the throttle plate will be partly open with the engine running fast. The obstruction presented by the throttle reduces the depression in the mixing chamber, and thence in the suction chamber, and the piston falls until the forces acting above and below the suction disc are in balance. The depression at the jet will be the same as before but as the jet has been raised the mixture will be weaker.

The high manifold depression will be sensed just downstream of the throttle plate at the auto-timing connection (on units so fitted) and transmitted to the distributor to advance the ignition timing, with beneficial effects on fuel economy.

At full throttle condition the throttle plate presents only slight obstruction, manifold depression is sensed in the suction chamber and the piston lifts to its maximum limit. The depression at the jet is as before but more fuel is drawn out due to the withdrawal of the tapered needle, which creates a greater annular area. The mixture strength remains constant.

7 Acceleration

A richer mixture is required for smooth acceleration. This is achieved by increasing the depression at the jet for any given piston position by retarding the rate at which the piston lifts, thus temporarily increasing the air velocity. The hydraulic damper, in conjunction with the oil filled piston rod, controls the rate at which the piston rises.

Note that the damper does not limit or reduce the height to which the piston rises, it merely controls the speed at which it rises, causing an immediately richer mixture to be delivered when the throttle is snapped open, which gradually weakens as the piston lifts to its correct position.

As the damper is only effective for upward movement of the piston, the piston falls immediately the throttle is closed, preventing an over-weak mixture being produced.

8 Special overhaul procedures

1 Refer to Chapter 5 Overhaul.
2 Gasket and overhaul packs – SU part numbers:

Carburettor model	Gasket pack	Overhaul pack
H1	*AUE 800*	*AUE 850*
H2	*AUE 800*	*AUE 850*
H4	*AUE 801*	*AUE 850*

9 Disassembly

1 Clean the outside of the carburettor thoroughly. Remove the banjo bolt, banjo union and fibre washers. Extract the filter and spring assembly from inside the float chamber lid.
2 Mark the relative position of the suction chamber to the body.
3 Remove the damper and its washer. Unscrew the chamber retaining screws. Lift straight off the chamber.
4 Remove the piston spring and washer (if fitted) and lift out the piston assembly carefully and empty out the damper oil from the piston rod.
5 Undo the needle locking screw and remove the needle. If the needle sticks in place, first tap it inwards and then pull it out; do not bend it.
6 Unhook the lever return spring. Remove the split pins and

9.1 Note how the filter sits in the spring

9.3 Lift the chamber fair and square away from the base

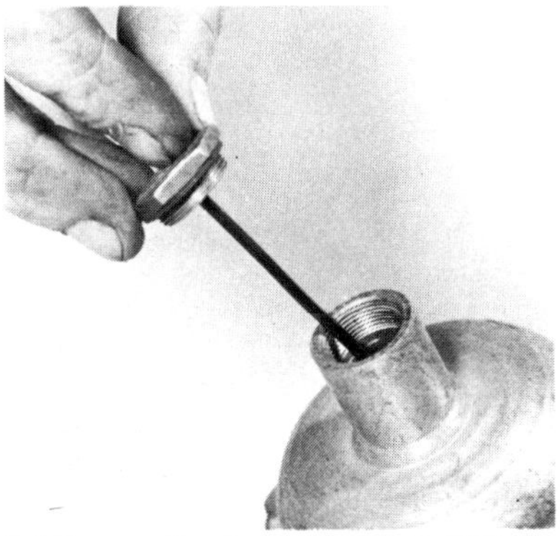

9.4 Remove the damper and then pour away the damper fluid

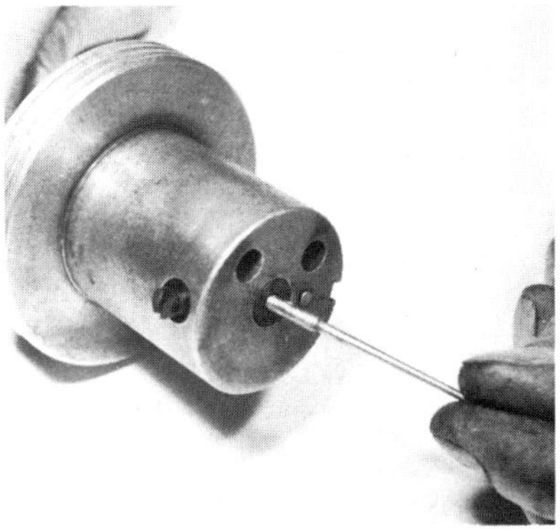

9.5 Do not tap the needle too hard should it stick; they bend easily

9.6 This is the standard arrangement of pins and springs

9.6a The clevis pins may not slip out too easily

9.8 Most threads of old carburettors will be stiff and fragile

9.10 Use the correct size of tool for all jobs; damage will occur if not

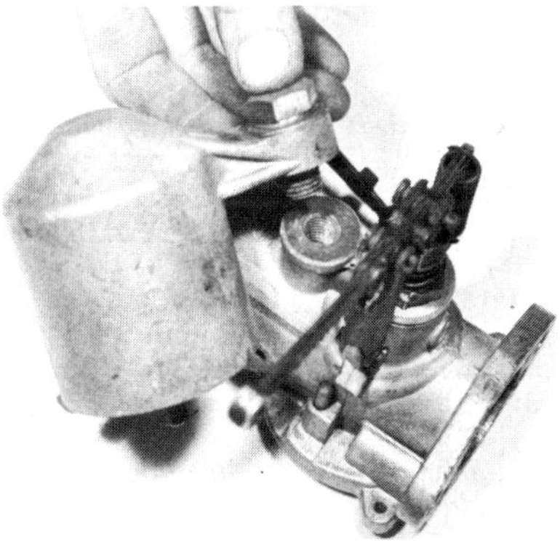

9.11 The levers are still in place in this photo to place the components better

clevis pins. Remove the fast idle cam pivot bolt. Note the positions of the aluminium spacing washer and the spring washer.
7 Undo the linkage assembly. Press in the piston lifting pin, extract the circlip from its groove and withdraw downwards the pin and its spring.
8 Withdraw the jet and take off the jet adjusting nut and spring.
9 Unscrew the jet locking nut and carefully withdraw the assembly. Lift off the upper jet bearing and copper washer. Extract the gland and brass gland washer.
10 Remove the gland spring and extract the lower jet bearing from the jet locking nut. Extract the gland and brass gland washer from the bearing. Do not touch the jet locking nut cork washer.
11 Remove the screw retaining the stay to the carburettor body (if fitted). Remove the fixing retaining the float chamber to the body. Record the positions of the fibre washers and the brass washer.
12 Mark the position of the lid to float chamber. Unscrew the central nut and remove the overflow pipe and washers; the stay, washer and cover cap, or the cover cap alone. Record the positions of the washers and other components. Lift off the lid, note the gasket. Remove the float.
13 Pull out the hinge pin for the hinged lever and detach the lever.
14 Lift out the needle from its seating and unscrew the seating from the lid using a spanner 0.338 in/8.58 mm across flats. Do not distort the seating.
15 Remove the two disc retaining screws. Twist the throttle and ease out the disc from its slot in the spindle. Pull out the spindle from the body.

9.12 The float chamber lid seal may stick

9.12a Invert the chamber to let the float fall out

9.13 Pull out the hinge pin carefully

9.14 The action of the float needle is all important

9.15 A dumpy screwdriver is best for this job

9.15a Wiggle out the butterfly, but do it carefully

16 Loosen the return spring clip bolt and remove the clip, spring, and return spring plate (if fitted). If a clamped operating lever is fitted, loosen the clamping bolt and remove the lever.

17 Close the throttle and record the position of the throttle disc.

18 For instructions on cleaning, inspection and repair refer to Part 1, Chapter 5 of this manual.

10 Assembly

Note: Ensure that all parts are clean and dry before assembly.

1 Fit the spindle to the body, ensuring that the fixed lever is in the correct position. Slide the throttle disc into the slot in the spindle and fit two new retaining screws. Do not tighten at this stage. Close the throttle when the disc will centre itself in the bore. Check visually that contact is made between the disc and the bore throughout its circumference. Tighten the screws and spread the split ends sufficiently to prevent the screws unscrewing.

2 Fit the jet assembly in the reverse order to disassembly, using new gland packings. Ensure that the washer is under the shoulder of the lower jet bearing, that the coned faces of the gland washers face toward the gland packings, and that the copper washer is fitted with the sharp edge towards the upper jet bearing. Fit the assembly to the body but do not tighten the jet locking nut at this stage.

3 Fit the needle to the piston. The relative position of the two parts is critical and may be either of two arrangements according to the contour of the needle at junction of the taper and the shank (ie square or radiused shoulder).

Fit and tighten a new locking screw.

Fit the piston assembly to the body taking care not to damage the needle. Fit the washer (if provided) and piston spring to the piston rod. Lightly oil (SAE 20W/50) the outside of the piston rod and fit the suction chamber. Fit and tighten the securing screws.

4 Remove the jet, jet locking nut, and spring. Refit the adjusting nut and screw it up as far as it will go. Refit the jet and ensure that it is in the correct relative position (check position of fork). With the jet locking nut loose, check that the underside of the piston is in contact with the top of the jet, which will protrude into the bore. The jet is now accurately centred. Tighten the jet locking nut, remove the jet and adjusting nut, fit the spring and refit the adjusting nut and jet. Fit the damper and washer but do not fill piston rod with oil at this stage.

5 Fit the float chamber valve seating to the float chamber. Fit needle, hinged lever and hinge pin. With the hinged lever resting on the seated needle check that a 0.44 in/11 mm diameter bar can be inserted between the forked lever and the lip of the float chamber lid. If there is a gap, or if the bar lifts the forked lever clear of the needle, adjust by bending the lever where shown.

6 Fit new float chamber lid gasket (do not use jointing compound), float (check that it is the correct way up) and float chamber lid. Ensure that the lid is in the position marked on disassembly. Fit the cover cap and central nut, or drainpipe, washers and nut (alternative detail). Use new fibre washers. Do not overtighten nut.

7 Fit the float chamber assembly to the carburettor body using new fibre washers or rubber grommets (alternative detail).

8 Insert the fuel inlet filter assembly (spring end innermost) and fit the banjo and bolt with new fibre washers. Note that the recessed face of the banjo fits toward the hexagon end of the bolt.

9 Fit the return spring plate, return spring and return spring clip to the throttle spindle. Tension the spring by turning the clip on the spindle and tighten the clip pinch-bolt. Fit the linkage assembly using new split pins. Ensure that the distance washer and double-coil spring washer are in their correct positions in relation to the fast-idle cam.

11 Tuning

Note: Refer to Part 1, Chapter 4 of this manual for notes on tuning, with particular reference to preliminary procedures.

10.1 A little drop of oil on the spindle will help assembly

10.1a Be sure before finally tightening

10.1b Make sure the throttle butterfly has a full range of movement

10.2 Do not forget the sealing washer before installing the jet

10.2a The complete assembly, but not yet installed

10.3 Go very carefully at this stage

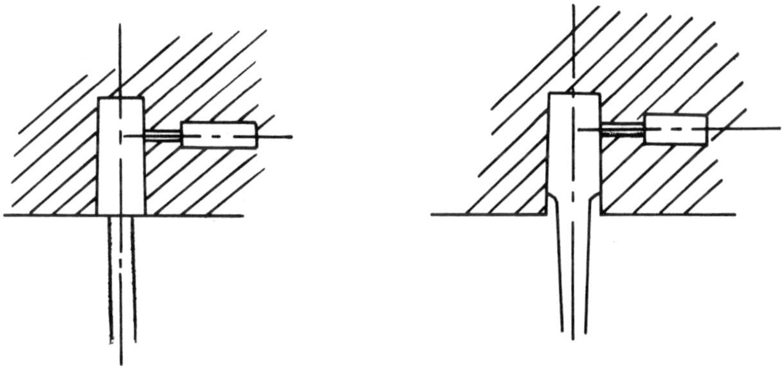

Fig. H5. See 10/3 - the two types of needle fitting

10.3a A certain amount of caution is still needed even here

10.4 Jet centering should never be ignored; it is essential for efficient running

10.5 A drill bit works best for this measurement

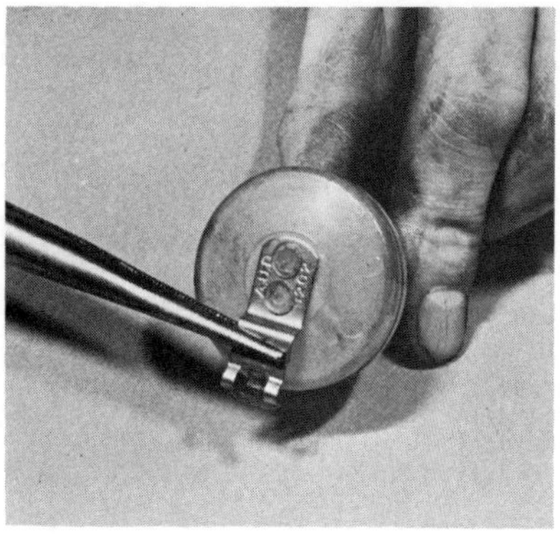

10.5a Bend the lever very gently, not too much, not too little

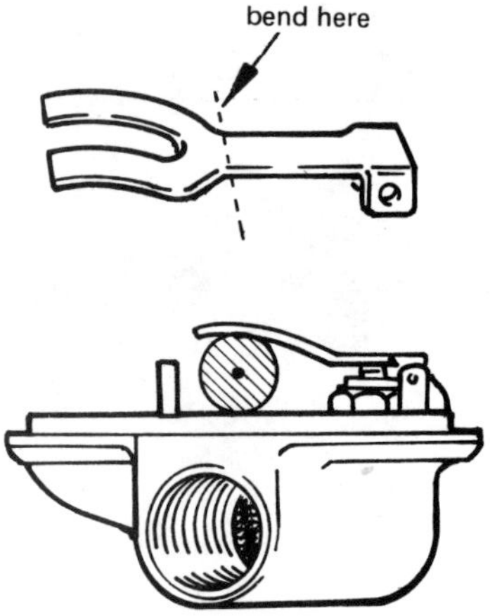

Fig. H6. Needle valve adjustment

10.6 Use a new lid gasket if possible

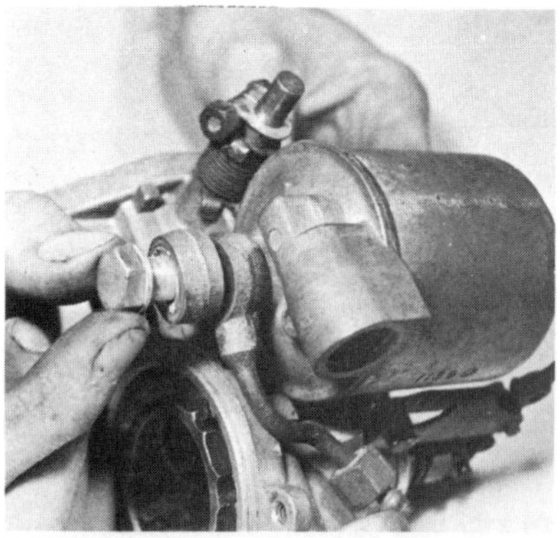

10.6a Never lose any of the proper washers when locating the fixing screw

10.7 Again position the float chamber back to its original position

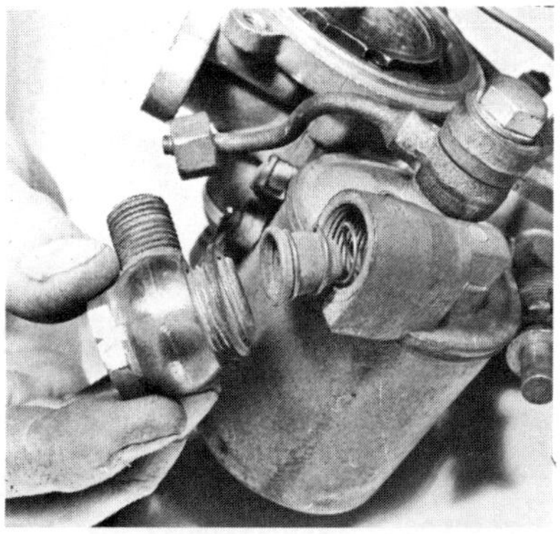

10.8 Cleanliness is all important in rebuilding carburettors

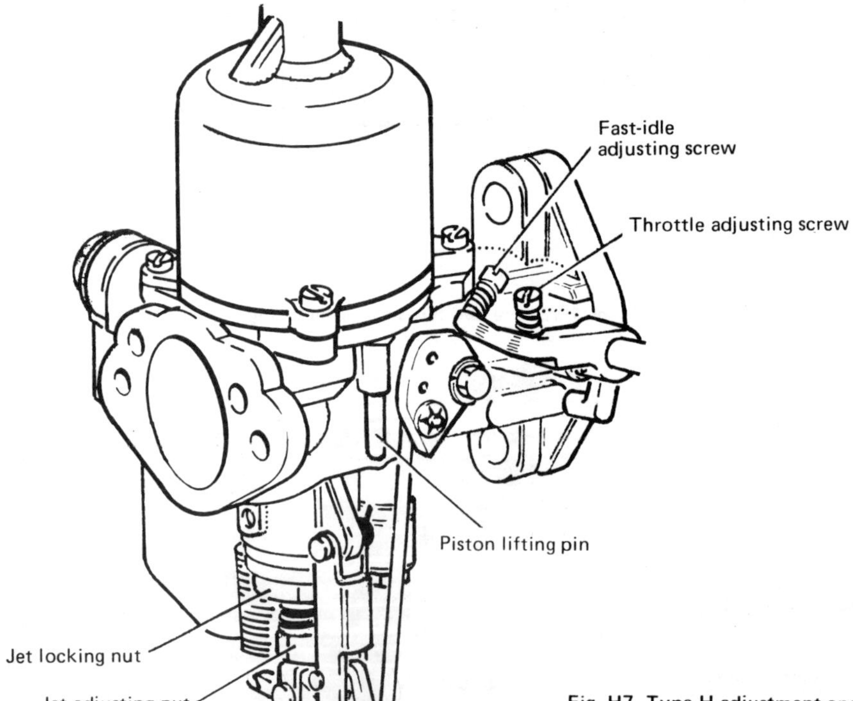

Fig. H7. Type H adjustment and tuning facilities

Single carburettor installations

1 Set throttle adjusting screw.

Warm up engine to working temperature. Switch off. Unscrew the throttle adjusting screw until it is just clear of its stop and the throttle is closed. Set throttle adjusting screw one-and-one half turns open. Ensure that fast-idle screw is clear of fast-idle cam.

2 Centre jet.

Mark for proper reassembly and remove piston and suction chamber unit. Disconnect the mixture control wire. Screw the jet adjusting nut until the jet is just flush with the bridge of the carburettor or fully up if this position cannot be obtained.

3 Set jet adjusting nut.

Replace the piston and suction chamber unit into original position. Check that the piston falls freely onto the bridge when the lifting pin is released. Turn down the jet adjusting nut two complete turns.

4 Set idling speed and mixture strength

Start engine and adjust throttle adjusting screw to give desired idling speed. Turn the jet adjusting nut up to weaken or down to richen until the fastest idling speed consistent with even running is obtained. Readjust the throttle adjusting screw to give correct idling if necessary.

5 Check mixture strength. (See Fig. H11 and H12).

The effect on exhaust smoke of mixture strength

(w)	Too weak	Colourless, irregular note, splashy misfire:
(c)	Correct	Regular and even note.
(r)	Too rich	Blackish, regular or rhythmical misfire.

Check for correct mixture by slowly pushing the lifting pin up about 0.031 in/0.8 mm). The graph illustrates the effect on engine rpm as the lifting pin raises the piston, indicating the mixture strength.

(r)	Rich mixture	rpm increase considerably
(c)	Correct mixture	rpm increase very slightly
(w)	Weak mixture	rpm immediately decrease

6 Connect and set mixture control wire.

Connect the mixture control wire and adjust with about 0.0625 in/1.6 mm free movement before it starts to pull on the jet lever. Pull the choke knob at the facia until the linkage is about to move the carburettor jet and adjust the fast-idle screw to give an engine speed of about 1000 rpm when the engine is hot.

7 Fill damper.

Finally top up the piston damper with thin engine oil grade SAE 20W/50 until the level is 0.5 in/13 mm above the top of the hollow piston rod. (On dust-proofed carburettors, identified by no vent hole in the damper cap, and a transverse hole drilled in the neck of the suction chamber) the oil level should be 0.5 in/13 mm below the top of the hollow piston rod.

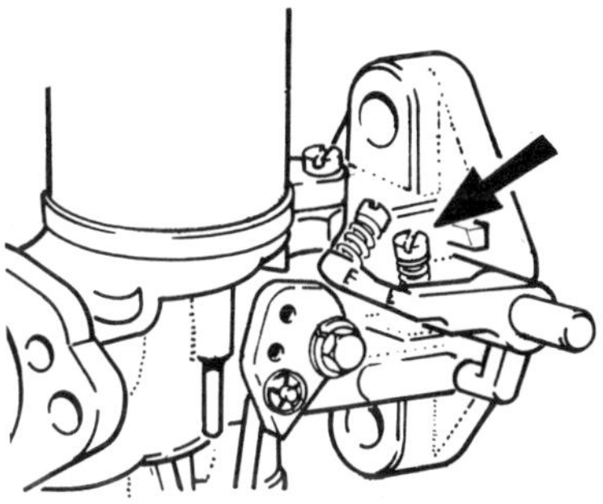

Fig. H8. Close up of the throttle adjusting screw

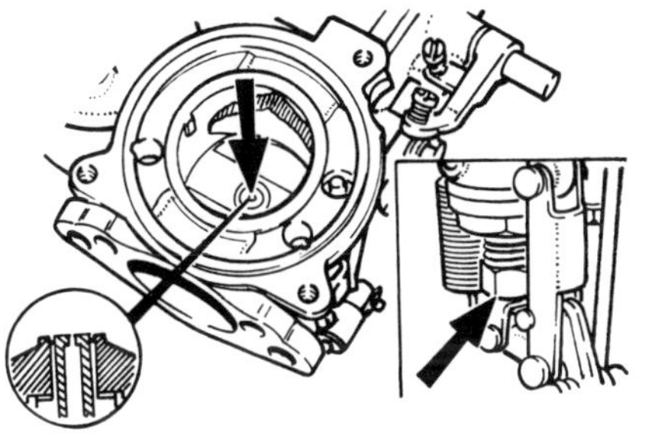

Fig. H9. Jet centering

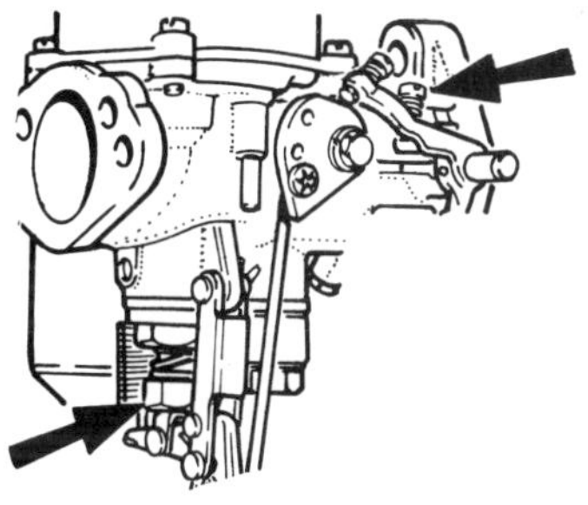

Fig. H10. Setting idling speed and mixture strength

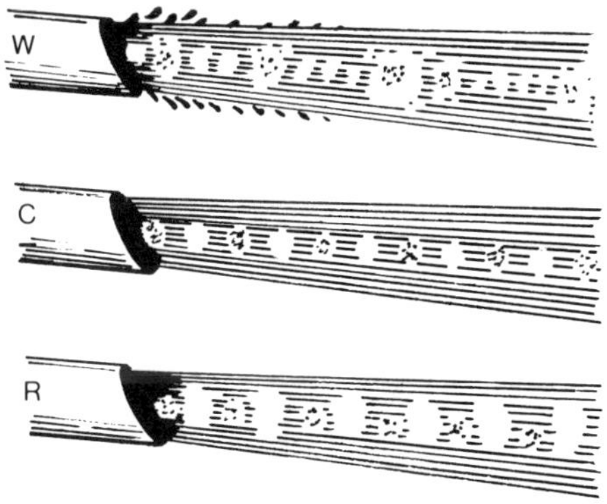

Fig. H11. Mixture strength as a function of exhaust smoke

w - too weak
c - just right
r - too rich

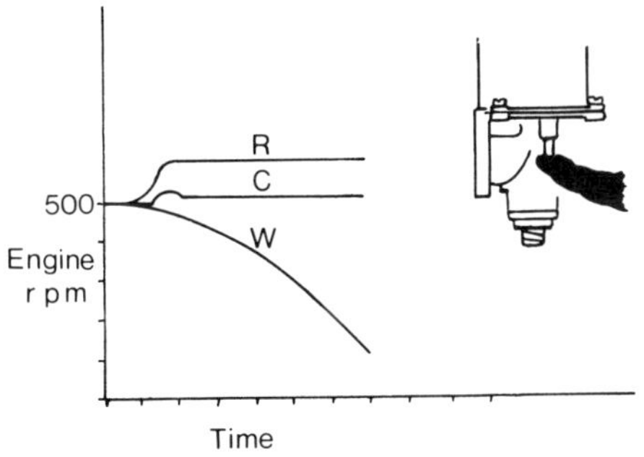

Fig. H12. Mixture strength on a graph

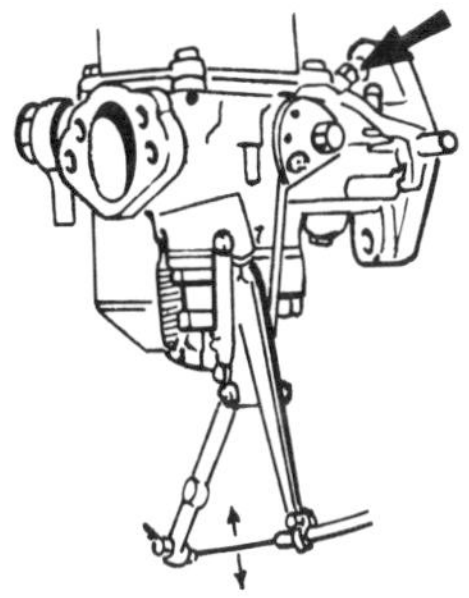

Fig. H13. Setting the choke

Multiple carburettor installations

Before the attempted tuning of any carburettor installation it is essential to make sure the valve clearances, points gap and spark plug gaps are correct. Successful tuning cannot take place if any one of these is 'out'. Multiple carburettor set-ups have their problems compounded if this is not strictly observed.

1 Remove the air cleaner(s) and check for throttle linkage adjustment, smoothness of action and that the dashpots are topped-up properly. *Note:* Whenever throttle adjusting screws are fitted they, and not the slow running valves, must be used to adjust the idling speed. Screw down the slow running valves (which must remain closed), and set the throttle adjusting screws one-and-one half turns open. Adjusting the idling speed with the throttle adjusting screw is dealt with later.

2 Slacken a clamping bolt on one of the throttle spindle interconnection couplings between the carburettors.

3 Disconnect the jet control interconnecting rod or cable (choke control) at the forked end.

4 Restart the engine and turn the slow running valve, or throttle adjusting screws, an equal amount on each carburettor to give the desired idling speed.

5 Compare the intensity of the intake hiss on each carburettor and alter the slow running valves or throttle adjusting screws, until the hiss is the same in each intake.

6 Turn the jet adjusting screw an equal amount on each carburettor, up to weaken or down to richen, until the fastest idling speed consistent and balanced with even running is obtained.

7 Readjust the slow running valves, if necessary.

8 Check the mixture by raising the lifting pin of one of the carburettors 0.03125 in/1 mm after free movement has been taken up. The graph illustrates the possible effect on engine rpm.

9 Repeat the operation on the other carburettor(s) and after adjustment re-check again as of course the carburettors are interdependent.

10 Fig. H11 shows the effect of mixture on the exhaust smoke.

11 Tighten the clamp bolt of the throttle spindle interconnections with the pin of the link pin lever resting against the edge of the pick-up lever hole. When forked levers are fitted, set the cranked levers so that the pin is ·006 in/·15 mm from the lower edge of the fork. This makes sure that both (all) carburettor throttles operate simultaneously.

12 Reconnect the jet control linkage so that the jet operating arms move simultaneously; if necessary, turn the fork end(s) and adjust the tension of (choke) the rod.

13 Reconnect the mixture control wire with about 0.0625 in/ 1.6 mm free movement before it starts to pull on the jet levers.

14 Pull the choke knob until the linkage is about to move the carburettor jet operating arms, and adjust the fast idle screws to give an engine speed of about 1000 rpm when hot.

15 Return the choke and check that there is a small clearance between the fast idle screws and the throttle stops.

16 Refit the air cleaners and re-check for correct mixture again.

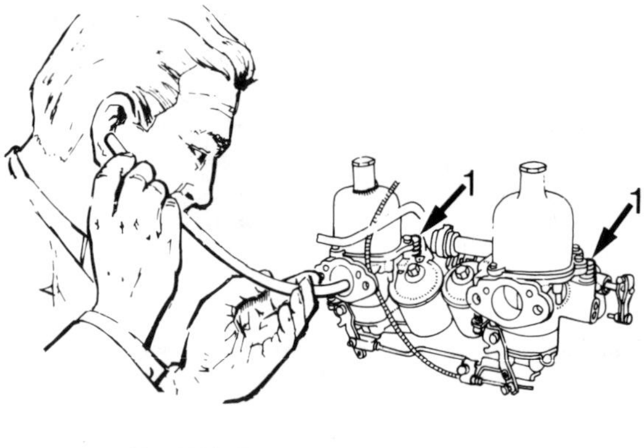

Fig. H14. The standard method of balancing two carburettors

1 Throttle adjusting screw

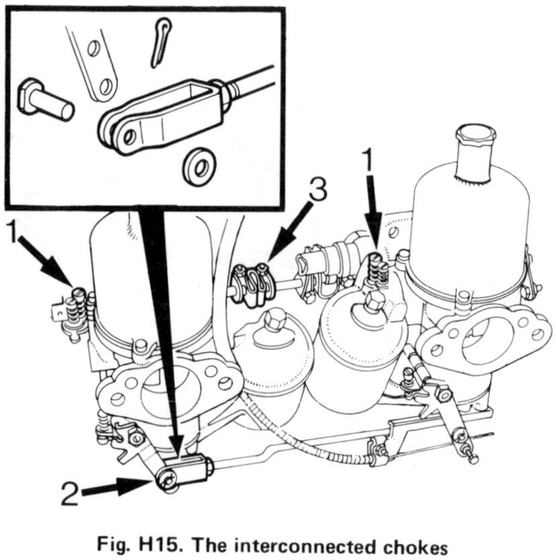

Fig. H15. The interconnected chokes

1 Throttle adjusting screw
2 Choke connection
3 Throttle spindle connection

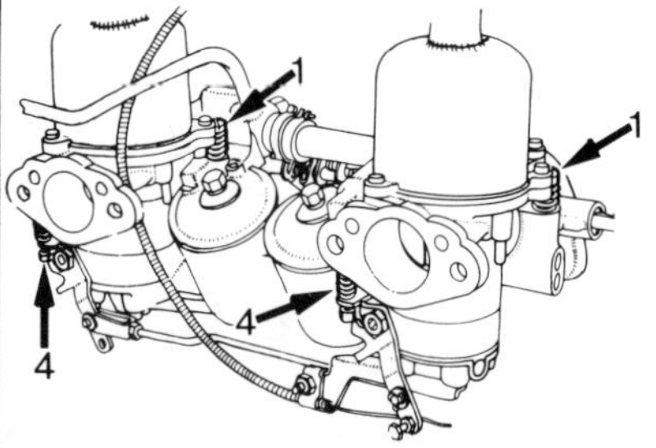

Fig. H16. Individual carburettor slow running adjustment

1 throttle adjusting screw
4 mixture screws

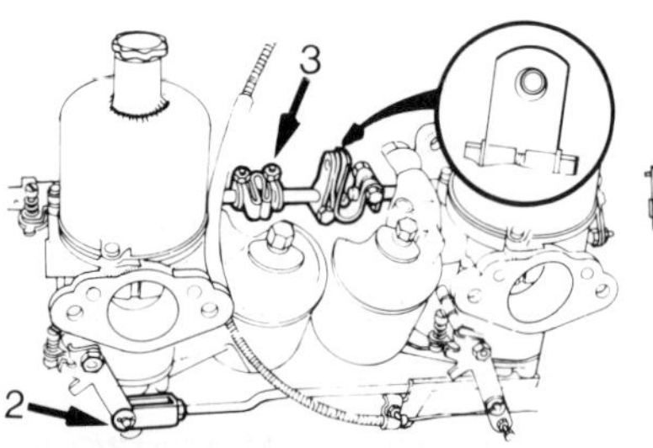

Fig. H17. Adjusting the throttle spindle connections

2 Choke connection
3 Throttle spindle connections

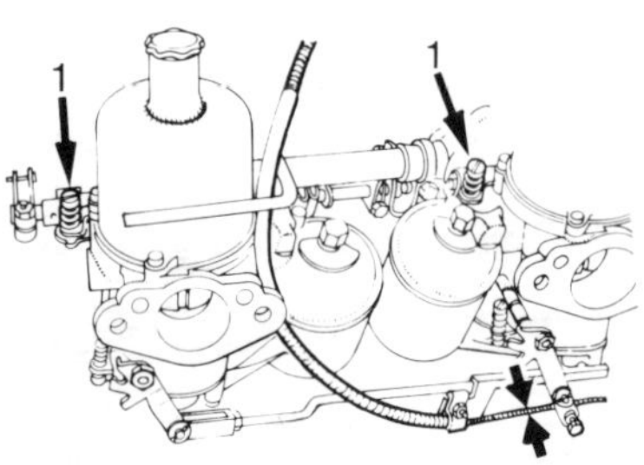

Fig. H18. Interconnection between the choke knob on the facia and the carburettors

1 Throttle adjusting screw

Type HD

Contents

1 Introduction

The Type HD (Diaphragm-jet) carburettor was introduced after the Type H. It was manufactured in larger sizes only — 1½ in, 1¾ in and 2 in, being models HD4, HD6 and HD8 respectively, and was therefore generally fitted to larger and heavier cars.

The characteristic features of the Type HD are the methods of metering fuel for idling, and of sealing the jet base.

2 Construction

The piston and suction chamber assembly is of the same general design as that of the Type H. The dashpot bore may be either dustproofed or non-dustproofed and may therefore have un-drilled or drilled piston damper caps.

The jet slides within a bearing secured to the underside of the body by the same method as in the Type H, and is surrounded by an open-ended cylindrical housing, cast integrally with the body. The housing is closed by a jet housing incorporating the jet actuating lever assembly hinged in integrally-cast lugs, and an extension of the float-chamber, the ports being secured to the body by four long screws. A flexible diaphragm (fixed to the base of the jet assembly) is clamped between the float-chamber extension and the underside of the jet housing and divides the housing cavity into two chambers. The upper chamber contains the jet actuating (lowering) mechanism and the lower chamber is in communication with the float chamber via a drilling in the extension, and is therefore flooded with fuel. The fuel flows up through the hollow jet to the orifice at the top.

A helical compression spring, fitted between a cup at the base of the jet assembly and a locating spigot in the float-chamber extension, loads the jet upwards.

All HD4 and HD6 units are fitted with throttle spindle sealing glands, which minimise inwards air leakage when the spindle bearings become worn. Some HD8 units are also so equipped, while others have no sealing glands and are fitted with replaceable PTFE bushes.

The glands have tapered faces and are fitted into conical

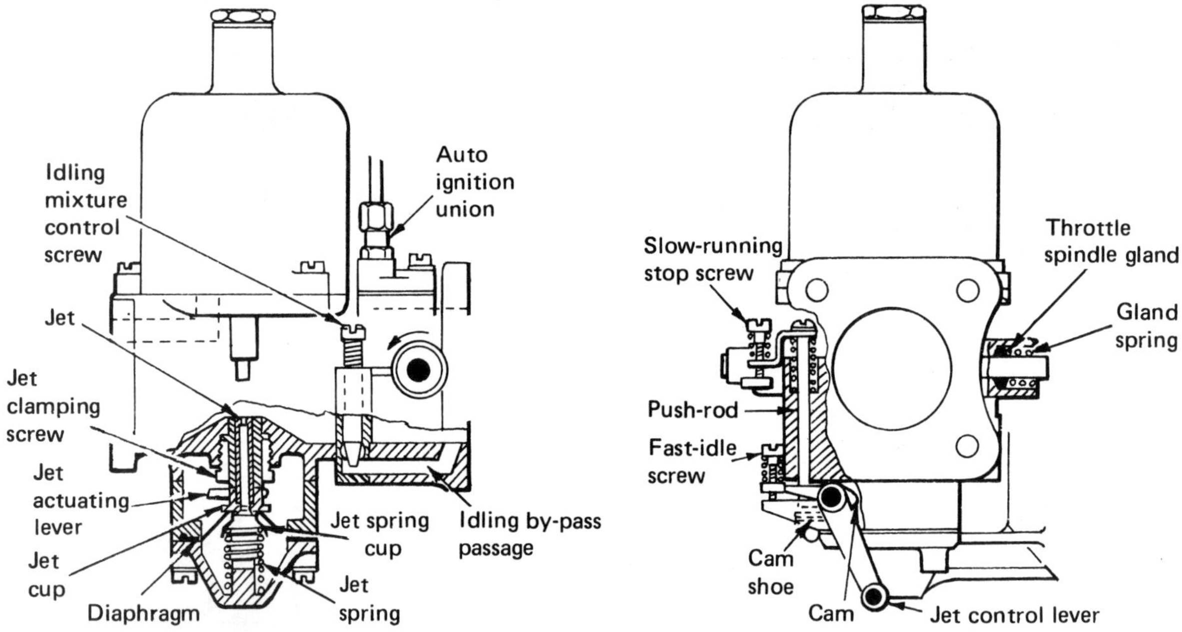

Fig. HD2. Cross sections through a Type HD carburettor

ended counterbores in the unit body. A helical compression spring, retained by a steel sleeve pressed into the end of the counterbore, compresses the gland so that the bore closes onto the throttle spindle and forms a seal. The glands do not require servicing and no provision is made for their easy removal.

PTFE bushes are inserted into counterbores in the unit body from inside the bore of the unit. They are retained in position by circular, spring clips fitted to the throttle spindle.

The underside of the body has cast ribs and there is a cast projection at each side of the main bore. The ribs are internally drilled to form a passage which by-passes the throttle plate when it is closed. One of the side projections houses a screwed needle valve, the end of which intercepts the drilled passage and functions as an idling mixture volume control valve. The other side projection has a vertical bore in which slides a cam rod, forming part of the cold start enrichment/throttle opening interconnection mechanism. The top end of the cam rod is fitted with a plate carrying an adjusting screw which bears on the throttle spindle lever. The bottom end is rivetted to an arm at the end of which is a roller. The roller bears on a cam at the end of the jet actuating lever spindle.

The float chamber assembly is of similar design to Type H, in that it houses a brass float sliding on a central spindle and the inlet valve is closed by a hinged lever mounted in the detachable float chamber top.

An ignition timing vacuum tapping is provided in the roof of the bore, the fine hole emerging at the edge of the throttle plate when in the closed position. The hole is bored through from a flat projection behind the suction chamber. Connection is made by an adapter plate with a stub pipe, secured to the body with two screws. A gasket is fitted between the body and the plate.

3 Jet assembly

The jet assembly consists of a brass jet tube, a flanged collar on which the jet actuating lever bears, a spring cup and a synthetic rubber diaphragm. The component parts are assembled by clamping and pressing and cannot be separated.

The diaphragm has a moulded circumferential rib on its upper surface, which locates in a similar annular groove in the underside of the jet housing and forms the sealing face. Four holes or cut-outs on the periphery accommodate the jet housing securing screws.

The jet size is identified by a code number stamped on the upper surface of the flanged collar.

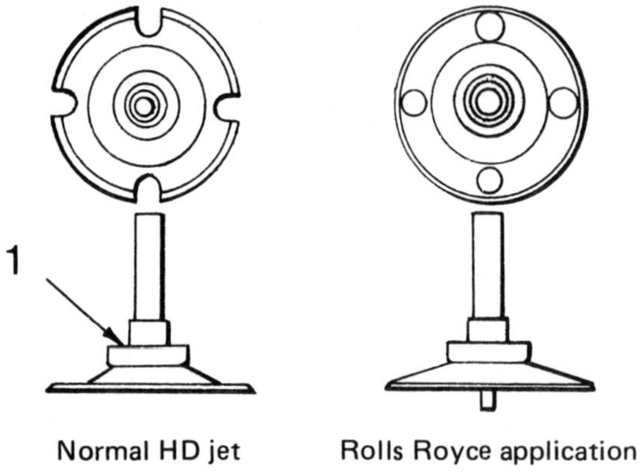

Fig. HD3. Type HD jets - standard and Rolls Royce

1 Indicates the jet size stamped on the jet.
Jet size 9 - 0.09 in, 1 - 0.01 in, 125 - 0.125 in

4 Operation

The operation of the unit for acceleration, full throttle and cruise conditions is similar to that of the Type H. The idling condition is different as the throttle disc is not used for idling mixture volume regulation.

When idling (engine hot), the throttle plate is fully closed in the carburettor bore. Mixture flows from the jet (upstream of the throttle) through the by-pass passage, past the idling mixture control screw and returns to the bore through a drilling downstream of the throttle plate. This gives more precise control of metering small quantities of mixture and eliminates the effects of irregular flow around worn throttle plates, bores and spindles.

Note that the idling mixture *strength* is still determined by the position of the jet, and that the idling mixture control screw only regulates the *volume* of mixture.

At cold-start condition, the jet is lowered (by operation of the jet control lever) and the throttle interconnection mechanism opens the throttle slightly to a pre-set amount (determined by adjustment of the slow-running stop screw). A rich mixture of increased volume is therefore supplied to the engine through both the main throttle valve and the by-pass passage.

As the engine warms through the jet is raised and the throttle is automatically closed by the interaction of the throttle interconnection mechanism, when the 'choke' control is pushed home.

Due to the construction of the linkage in the jet housing it is possible to obtain a fast-idle condition by pulling out the 'choke' control slightly from its fully closed position. This has the effect of opening the throttle before the jet actuating lever starts to lower the jet. Mixture at normal strength and at increased volume is therefore supplied which is correct for fast running of a hot engine.

5 Special overhaul procedures

1 Refer to Part 1, Chapter 5 Overhaul.
2 Gasket and overhaul packs – SU Part numbers

Carburettor model	Gasket pack	Overhaul pack
HD4	*AUE 805*	*AUE 855*
HD6	*AUE 805*	*AUE 855*
HD8	*AUE 806*	*AUE 857*

6 Disassembly

1 Clean the outside of the carburettor thoroughly.
2 Undo and remove the damper and washer.
3 Remove the suction chamber retaining screws and remove the chamber straight up.
4 Lift off the piston spring. Carefully lift out the piston and needle assembly. Empty out the damper oil from the piston rod.
5 Remove the needle locking screw and withdraw the needle. If it sticks in place tap the needle inwards first and then pull outwards. Do not bend it.
6 Undo the plate retaining screw and lift off the plate and spring. Withdraw the cam rod assembly.
7 Record the positions of the float chamber, jet housing, and carburettor body. Unscrew the float chamber screws, holding the float chamber against the pressure of the jet spring. Carefully detach the float chamber.
8 Lift out the jet spring, Mark the jet diaphragm opposite one of the screw holes in the jet housing and withdraw the jet assembly and jet housing.
9 Slacken and remove the jet locking nut together with the jet bearing with the appropriate spanner.

6.3 Check for metal corrosion whenever dismantling a carburettor

6.4 At this stage this carburettor still looks in reasonable condition

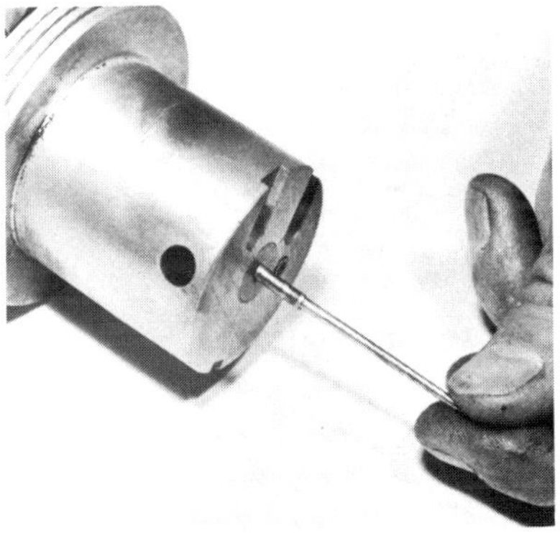

6.5 The needle is located in the same way as the Type H

6.5a Once again a dumpy screwdriver is best used for this job

6.7 Release each screw, then undo them all one after the other

6.7a Note the spring and gasket

10 Unscrew the banjo bolt from the float chamber and remove the bolt, banjo, and fibre washers. Pull out the filter and spring assembly from the float chamber lid inlet.
11 Record the positions of the float chamber and lid. Remove the central nut retaining the float chamber lid together with the overflow tube banjo and fibre washer, or cover cap, as fitted.
12 Remove the lid and gasket. Pull out the float lever hinge pin from the serrations end. Detach the lever.
13 Extract the float needle from its seating and unscrew the seating from the lid using a spanner 0.338 in/8.58 mm across flats. Do not distort the seating.
14 Remove the float from the chamber.
15 Shut the throttle and record the positions of the throttle disc and the carburettor flange.
16 Loosen and remove the disc retaining screws. The ends of the screws will be split to retain them securely - close the split before attempting to remove the screws.
17 Withdraw the disc from its slot in the throttle spindle. The disc is oval and can jam. Take care to make sure it does not.
18 Slide out the spindle from its bearings.
19 Do not remove the throttle spindle sealing glands. *Note:* Some Type HD8 carburettors are fitted with plastic spindle bushes and have no spindle sealing glands.
20 Loosen and remove the slow running valve complete with spring, seal, and brass washer.
21 Remove the two screws and washers holding the vacuum ignition take-off plate and union. Lift off the plate and gasket.
22 Remove the piston lifting pin by extracting the circlip from its groove with the pin pressed upwards. Withdraw the pin downwards.
23 For instructions on cleaning, inspection and repair refer to Part 1, Chapter 5 of this manual.

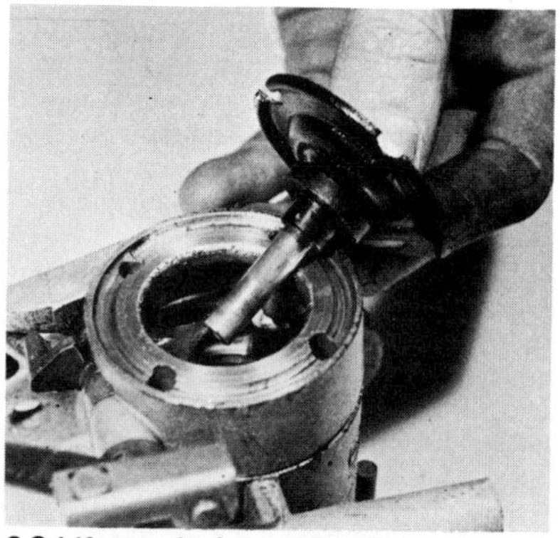

6.8 Lift out the jet and continue to look at the condition of the components

6.8a Jet housing removal

6.8b All the components of the jet housing, assembled but off the carburettor

6.9 Use a box spanner on the jet locking nut

6.16 Close the split in the screws before attempting to undo them

6.17 This carburettor has obviously had some usage

7 Assembly

1 Fit the spindle in its bearings.
2 Fit the throttle disc to the slot in the throttle spindle in the position marked previously (on disassembly). Note that the countersunk ends of the screw holes in the spindle must face outwards, ie towards the intake flange of the carburettor body. Insert two new retaining screws but do not tighten. Adjust the disc in the slot until it closes fully in the bore of the carburettor, and tighten the screws. Spread the split ends of the screws just enough to prevent them turning.
3 Fit the slow-running valve assembly taking care that the brass washer is fitted with its concave face toward the seal.
4 Fit gasket and vacuum ignition take-off plate, with screws and new shakeproof washers. Do not use jointing compound on the gasket.
5 Fit the piston lifting pin, spring, new rubber washer, plain washer and circlip.
6 Fit the needle seating and needle to the float chamber lid.
7 Fit the float lever and press in float lever hinge pin to secure. Check the float level as shown in Fig. H6 and adjust if necessary to achieve the dimension given by bending the lever at the point shown.
8 Refit the float to the float chamber.
9 Fit the gasket to the lid and replace the lid on the chamber in the same position as recorded on dismantling. Fit the fibre washer, overflow tube banjo, plain washer, and nut or cover cap and nut, as applicable. Do not overtighten the nut.
10 Fit the filter to the lid inlet, spring end first. Fit the banjo,

7.1 Use both hands to draw out the spindle, one on the body and one on the spindle

7.2 Check for full movement and particularly for full closure

7.3 Take great care with the sealing washer

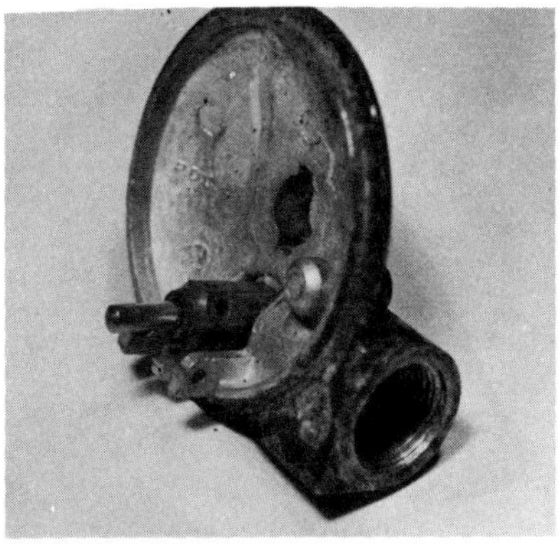

7.6 If in any doubt always replace the needle valve

7.7 Lever arm movement must be correct

7.8 For all practical purposes floats cannot be repaired

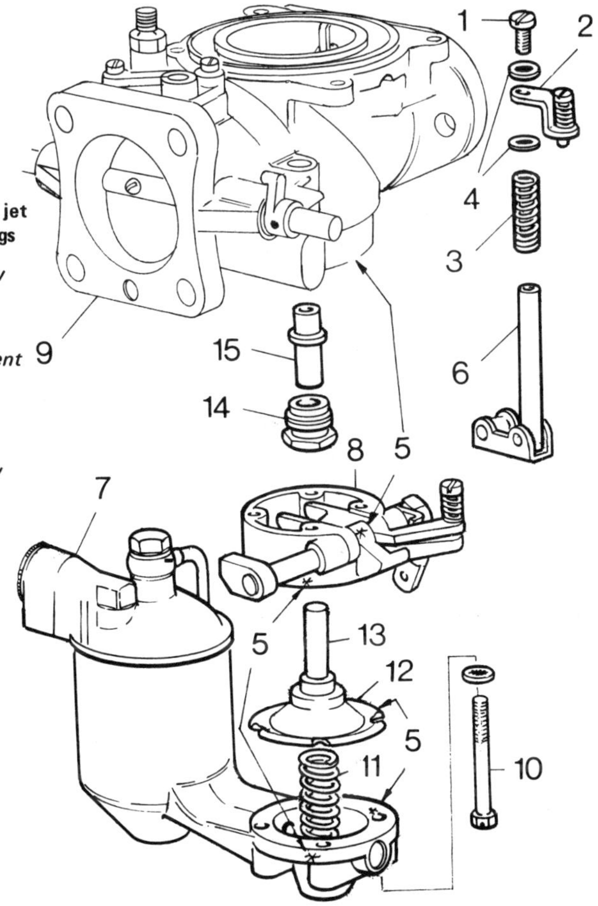

Fig. HD4. The type HD jet and float chamber fixings

1 Plate retaining screw
2 Plate
3 Spring
4 Shakeproof washer
5 Marks for replacement
6 Cam rod assembly
7 Float-chamber
8 Jet housing
9 Carburetter body
10 Float-chamber screw
11 Jet spring
12 Jet diaphragm
13 Jet assembly
14 Jet locking nut
15 Jet bearing

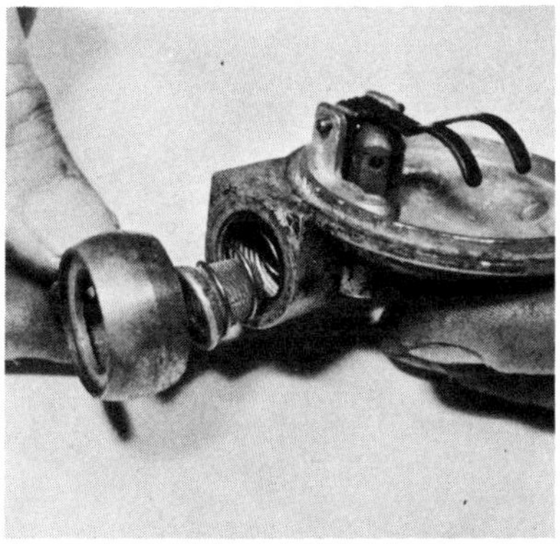

7.9 Never leave out the filter; it could save your engine

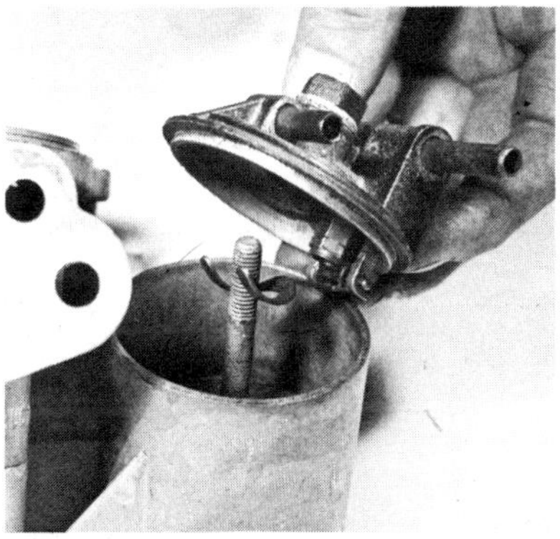

7.9a Try to make sure everything is cleaner even than this

fibre washers, and banjo bolt. The recessed flat of the banjo must face the hexagon of the bolt.

11 Fit the damper assembly and washer. Seal the transfer holes in the piston assembly with plasticine and fit the assembly to the suction chamber. Invert the assembly and allow the suction chamber to fall from the piston. Check the time this takes. It should be between 5 and 7 seconds. If the time taken exceeds that time quoted the cause will be thick oil on the piston rod or an oil film on the piston or inside the suction chamber. Remove the oil from the points suggested and re-test.

12 Fit the needle to the piston assembly. The lower edge of the groove must be level with the lower face of the piston rod. Fit a new needle locking screw and tighten. Invert the suction chamber and turn the piston assembly inside it to check for concentricity of the needlé.

13 Check the piston key for security in the carburettor body. Fit the piston assembly to the body and replace the piston spring on the piston rod. Fit the suction chamber and retaining screws. Tighten the screws evenly.

14 Fit the jet bearing and jet locking nut. Leave the nut with just enough slack to allow the bearing to be moved from side to side.

15 Fit the jet assembly to the bearing in the same position as recorded on dismantling. Centralise the jet.

16 Remove the jet and refit the jet housing, jet, jet spring and float chamber in the same relative positions as recorded on dismantling. Fit and tighten the securing screws evenly.

17 Fit the cam rod assembly and fit the spring, plate, and plate retaining screw with a shakeproof washer on either side of the plate. Make sure the plate is positioned so that its adjustment screw strikes squarely on the lug of the throttle spindle operating arm.

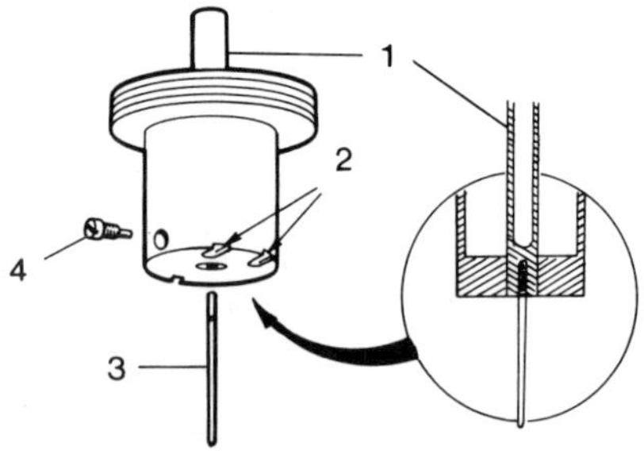

Fig. HD5. Piston needle fixing

1 Piston rod
2 Transfer holes
3 Needle
4 Locking screw

7.13 Make sure the piston key is where it should be

7.15 Jet centralising is as essential as any other carburettor adjustment

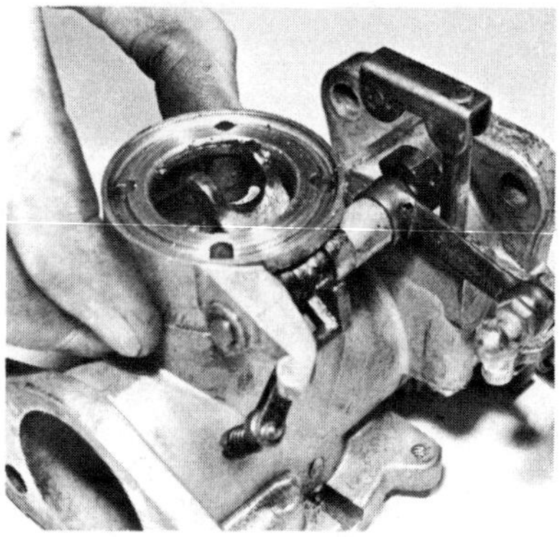

7.16 Align everything first; then press together

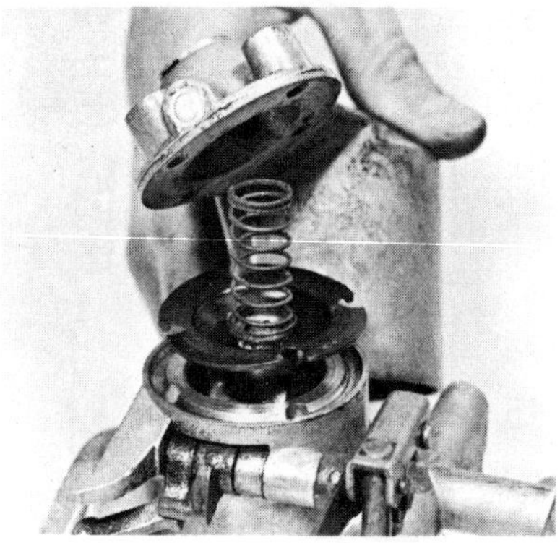

7.16a Jet, spring and float chamber body

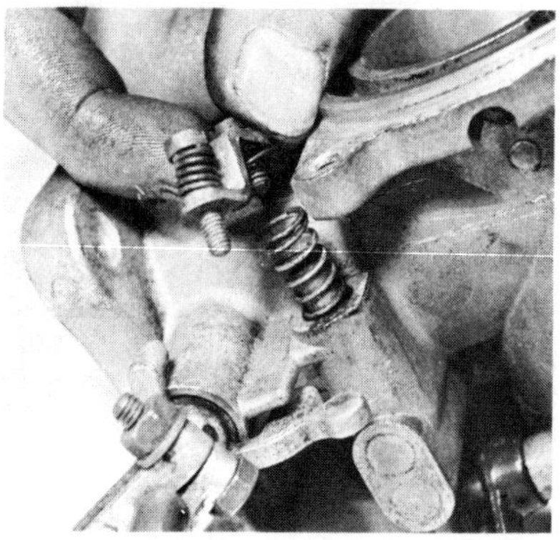

7.17 Make sure the order and completeness of the components is really right

8 Tuning

Note: Refer to Part 1, Chapter 4 of this manual for notes on tuning, with particular reference to preliminary procedures.

Single carburettor installations

1 Set fast-idle adjusting screw and slow-running valve.
Run the engine up to a normal running temperature. Switch off the engine. Unscrew the fast-idle adjusting screw to clear the throttle stop with the throttle closed. Screw down the slow-running valve onto its seating, then unscrew it three-and-one half turns.

2 Centre jet.
Remove the piston and suction chamber unit. Turn the jet adjusting screw until the jet is flush with the bridge of the carburettor.

3 Set jet adjusting screw.
Replace the piston and suction chamber unit. Check that the piston falls freely onto the bridge when the lifting pin is released. Lower the jet by turning the jet adjusting screw down two-and-one half turns.

4 Set idling speed and mixture strength.
Restart the engine and adjust the slow-running valve to give the desired idling speed. Turn the jet adjusting screw, up to weaken or down to richen, until the fastest idling speed consistent with even running is obtained. Re-adjust the slow-running valve to give correct idling if necessary.

5 Check mixture strength (See Fig. H11 and H12).
The effect on exhaust smoke of mixture strength.

(w)	Too weak	Colourless, irregular note, splashy misfire.
(c)	Correct	Regular and even note.
(r)	Too rich	Blackish, regular or rhythmical misfire.

Check for correct mixture by slowly pushing the lifting pin up about 0.031 in/0.8 mm. The graph illustrates the effect on engine rpm and indicated mixture strength as the piston is raised.

(r)	Rich mixture	rpm increase considerably.
(c)	Correct mixture	rpm increase very slightly.
(w)	Weak mixture	rpm immediately decrease.

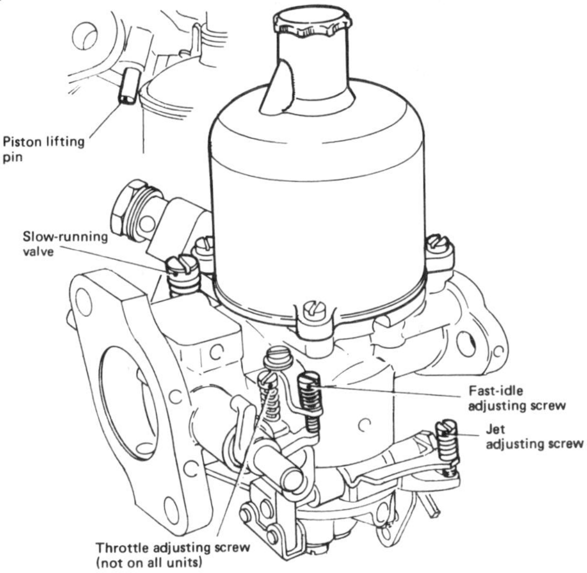

Fig. HD6. Type HD adjustment and tuning facilities

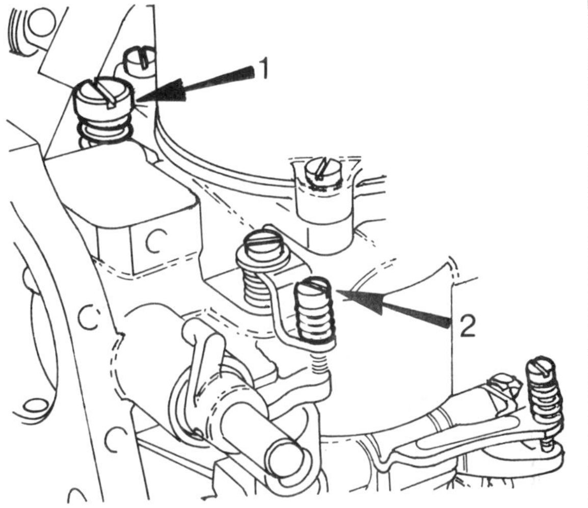

Fig. HD7. Basic tuning

1 Slow running valve
2 Fast idle adjusting screw

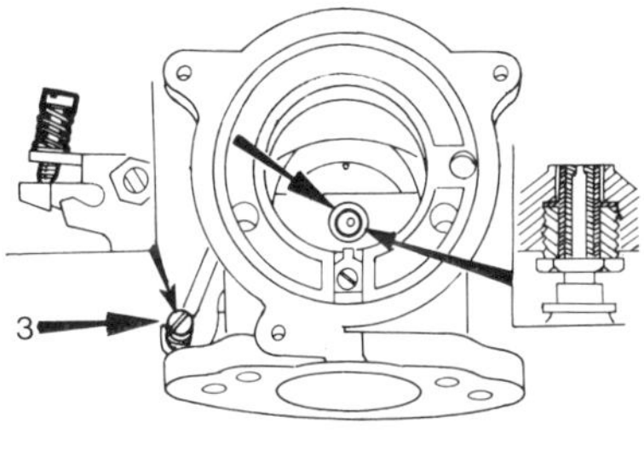

Fig. HD8. Jet centring

3 Jet adjusting screw

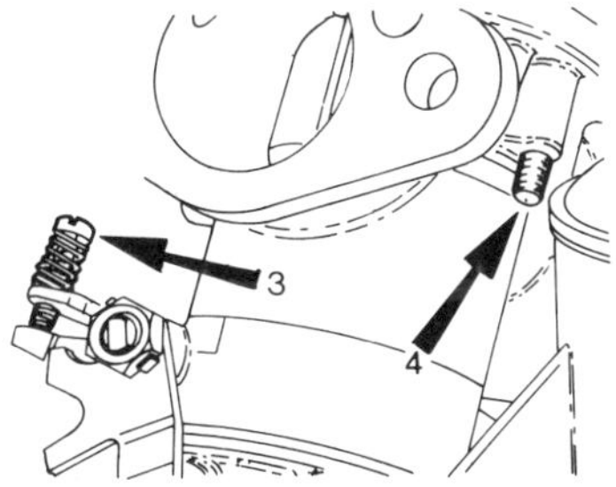

Fig. HD9. Jet adjusting screw setting

3 Jet adjusting screw
4 lifting pin

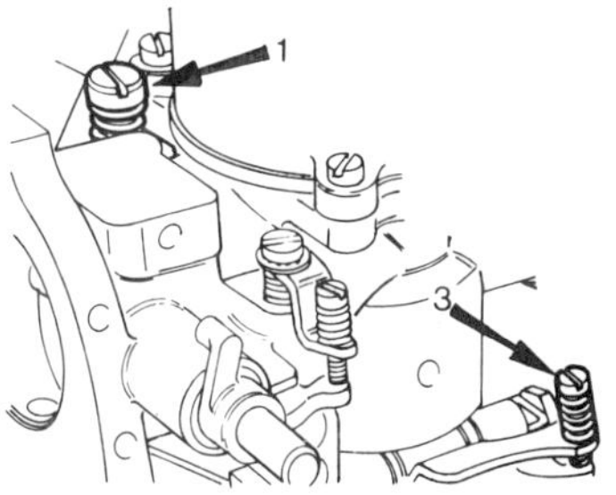

Fig. HD10. Final tuning

1 Slow running valve
3 Jet adjusting screw

6 Connect and set mixture control wire.
Connect the mixture control wire with about 0.0625 in/1.6 mm free movement before it starts to pull on the jet lever. Pull the choke knob at the facia until the linkage is about to move the carburettor jet operating arm and adjust the fast-idle screw to give an engine speed of about 1000 rpm when the engine is hot. Return the choke knob and check that there is some clearance between the fast-idle screw and the throttle stop.
7 Fill damper.
Finally top up the piston damper with thin engine oil grade SAE 20W/50 until the level is 0.5 in/13 mm below the top of the hollow piston rod. *Note:* On non-dustproofed carburettors, identified by a vent hole in the piston damper top, the oil level should be 0.5 in/13 mm above the top of the hollow piston rod.

Multiple carburettor installations
See this section for the Type H carburettor.

Fig. HD11. Choke setting

2 fast idle adjusting screw

Type HS

Contents

1 Introduction

The Type HS carburettor was introduced in the late 1950s and came into widespread use in the early 1960s. It does not supersede Types H or HD and was produced together with these units. It is probably the most popular type of SU carburettor in service at the present time.

The Type HS is basically a development of the Type H unit. The main improvement is in the jet assembly which was designed to eliminate the troublesome gland seals of the previous carburettor, and it is this feature by which it is easily identified. The float chamber is also of a different design and incorporates a float moulded from a plastic material instead of the previous brass type.

2 Jet assembly

Petrol from the float chamber flows direct to the base of the jet through a plastic tube. The tube is flexible which allows the jet to move down in its holder to give the necessary degree of enrichment for starting. This dispenses with the necessity of a petrol filled annulus surrounding a drilled jet tube, (as in Type H) and the attendant seals or glands at the base of the annulus to allow an extension of the tube to project through for external operation.

The jet assembly consists of an accurately machined brass tube (the jet), a flexible feed pipe, and a plastic moulding into which both items are inserted. They are secured by contraction of the moulding by means of conical metal collars pressed on. Short, brass, reinforcing sleeves are fitted inside each end of the flexible pipe to prevent it being crushed.

The jet holder is a brass tube, in which the jet is a sliding fit, externally flanged at the top and threaded at the bottom. It is held to the underside of the carburettor body by a large, steel nut which clamps the flange against the end of a counterbored hole in the body. The hole in the nut is considerably larger than the external diameter of the jet holder, thus permitting sideways adjustment of the jet assembly to centre it relative to the needle.

A helical compression spring and long nut are fitted to the threaded portion at the bottom of the jet holder. The underside of the nut is in contact with the metal collar on the jet assembly moulding, thus the vertical position of the jet is controlled by screwing the nut up or down. The jet is held in contact with the nut by spring loading on the external (choke) linkage.

The jet may be any of three sizes according to the size and application of the carburettor. The size denotes the diameter of the bore and is 0.09 in, 0.10 in or 0.125 in and is identified by the number of machined grooves at the top of the jet, being none, one and two, respectively.

If replacing a jet ensure that the replacement has the same number of grooves, the same colour of jet head (the plastic moulding), and the same number (one or two) and colour of sleeves as the original one. In the absence of confirmation from the manufacturers of SU carburettors, only by adopting this procedure can you be sure that the jet is the same. The identity of the jet is dictated by the following:

a) Angle of float chamber
b) Carburettor size
c) Interconnection (right- or left-hand)
d) Jet size

3 Float chamber assembly

The float chamber is of different design (from the previous Types H and HD) due to the different design of jet assembly, and a simplification and improvement in construction and operation.

The chamber is attached to the carburettor body by a single bolt screwed into a tapped boss cast on the side of the chamber. This enables the chamber to be rotated, within limits, to suit either horizontal or semi-downdraught installation, thereby eliminating the need for different castings. An adapter bush with locating tongues interposed between the chamber and the body ensures correct orientation for any particular installation. The fuel outlet from the chamber is arranged so that there is a sediment-well surrounding it. Where the outlet emerges from the chamber, the boss is tapped to accept the union nut of the jet connection pipe.

The detachable float chamber lid incorporates lugs which carry the float lever hinge pin. The fuel inlet is a stub (brass pipe) for connection to a push-on flexible pipe, and the vent is a horizontal drilling immediately below it, the outlet being shielded by a baffle plate to prevent the entry of dirt and to prevent fuel being ejected forcibly from it, should the inlet valve stick open.

The inlet valve may be either the brass type with a steel needle, as on Types H and HD, or may be a plastic-bodied, spring-loaded type. This latter type was introduced on the Type HS to overcome the effects of engine vibration which tends to affect the seating of the inlet valve.

To further assist in reducing the effects of vibration, the float chamber may be flexibly mounted by the inclusion of rubber bushes between it and the carburettor body. The float is formed by two plastic mouldings and the steel lever is integral with the top part.

4 Operation

The operation of the carburettor for cold-start, idling, acceleration and cruise conditions is the same as for the Type H.

5 Special overhaul procedures

1 Refer to Chapter 5 Overhaul.
2 Gasket and overhaul packs – SU Part numbers

Carburettor model	Gasket pack	Overhaul pack
HS2	*AUE 810*	*AUE 860*
HS4	*AUE 811*	*AUE 862*
HS6	*AUE 812*	*AUE 868*
HS8	*AUE 813*	

6 Disassembly

1 Take off the baffle plate from the inlet nozzle and thoroughly clean the outside of the carburettor.
2 Record the relative positions of the suction chamber and the carburettor body.
3 Take out the damper and its washer. Unscrew the chamber retaining screws and lift off the chamber without tilting it.
4 Take off the piston spring and washer (if fitted).
5 Lift out the piston assembly carefully and empty the oil from the piston rod.
6 Undo the needle locking screw and withdraw the needle. If it sticks, tap the needle inwards first and then pull outwards. Do not bend the needle.
7 If a piston lifting pin with an external spring is fitted, remove

Fig. HS2. Types of type HS jet assembly

A *early HS2* B *intermediate HS2*
C *latest HS2* D *HS8*

1 *brass*
2 *plastic*
3a *0.090" Jet (no groove)*
3b *0.10:: Jet (One groove)*
3c *0.125" Jet (two grooves)*
4 *brass inner tube*
5 *rubber washer*
6 *brass washer*
7 *identification sleeves*
8 *black plastic semi downdraught*
9 *red plastic horizontal*
10 *moulded swan neck*

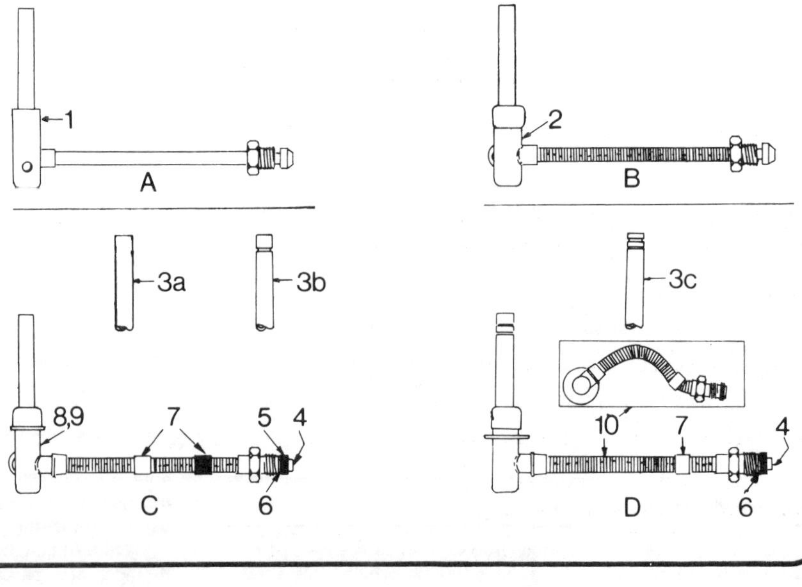

6.3 This shows the components in order

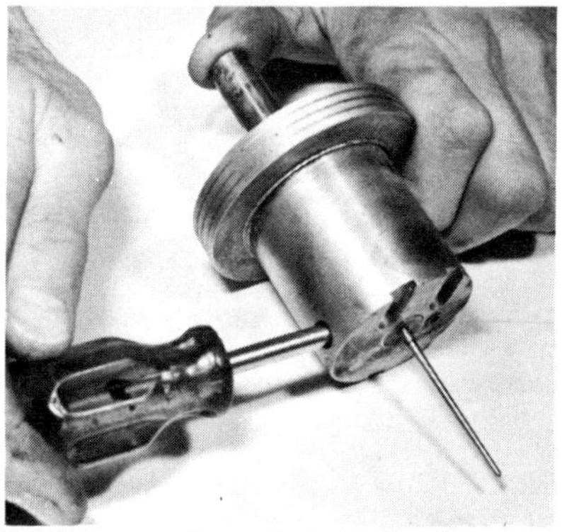

6.5 The correct way to hold the piston for needle removal

6.6 The damper rod complete

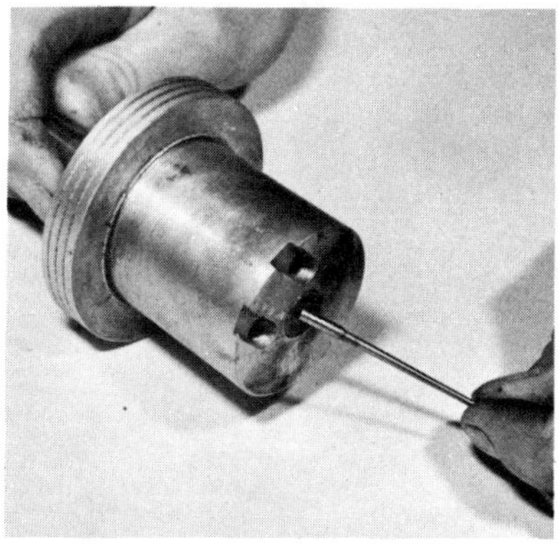

6.6a Pull firmly but in a straight line

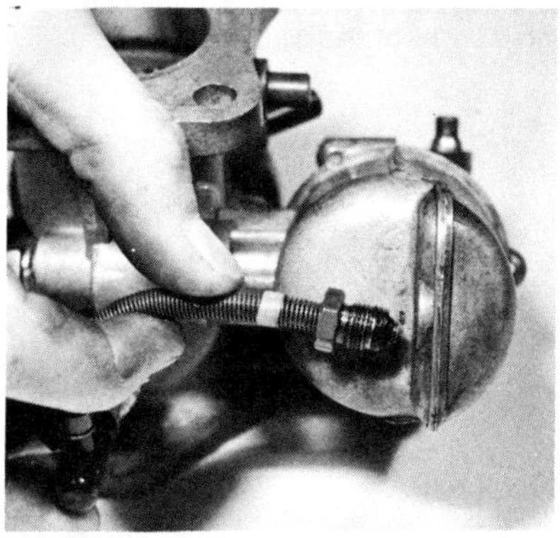

6.10 Always use the correct open ended spanner for this task

6.10a Now pull out the jet with great care

the spring retaining circlip and spring, then push the lifting pin upwards to remove it. For the concealed spring type, press the pin upwards, detach the circlip, and withdraw downwards the pin and spring.

8 Support the moulded base of the jet and loosen the screw retaining the jet pick-up link.

9 Relieve the tension of the pick-up lever return spring and remove screw and brass bush (if fitted).

10 Undo the brass sleeve nut retaining the flexible jet tube to the float chamber and take out the jet assembly from the carburettor body. Note the gland, washer, and ferrule, at the end of the jet tube.

11 Take off the jet adjusting nut and screw. Undo the jet locking nut and detach the nut and jet bearing. Withdraw the bearing from the nut.

12 Record the seating points of the two ends of the pick-up lever return spring. Unscrew the lever pivot bolt together with its double-coil spring washer, or spacer. Take off the lever assembly and return spring.

13 Record the seating of the two ends of the cam lever spring and push out the pivot bolt tube or tubes. Take care not to lose the spring. Lift off the cam lever, noting the washer between the two levers.

14 Loosen and take off the bolt retaining the float chamber to the carburettor body. Note the sequence with flexibly mounted float chambers.

15 Record the location of the float chamber lid. Undo the lid retaining screws and take off the lid and its gasket, complete with float assembly.

16 Pull out the float hinge pin from the serrations end and detach the float.

17 Extract the float needle from its seating and unscrew the seating from the lid, using a spanner 0.338 in/8.58 mm across flats. Take care not to harm any components.

18 Shut the throttle and record the relative positions of the throttle disc and the carburettor flange.

19 Unscrew the two disc retaining screws. Open and ease out the disc from its slot in the throttle spindle. The disc is not round

6.11 This is a newer less worn carburettor as can be seen by the clean threads

6.12 It is essential to record the setting prior to disassembly

6.13 Note the washers; this is your guide to reassembly

6.14 For clarity we have left on the other components of this carburettor

6.15 Do not forget to mark the lid for correct refitment

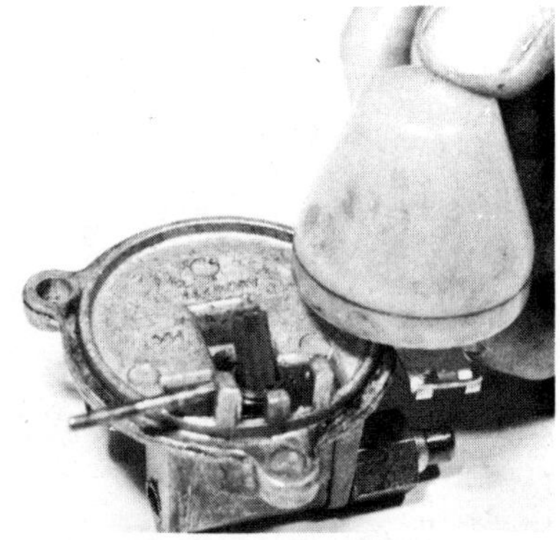

6.16 Note the end from which the hinge pin is withdrawn

6.19 Loosen both screws before withdrawing one

6.19a The butterflies are just out-of-round

but oval and will jam if care is not taken.
20 Bend back the tabs of the washer securing the spindle nut. Note the location of the lever arm in relation to the spindle and carburettor body; remove the nut and arm.
21 For instructions on cleaning, inspection and repair, refer to Part 1, Chapter 5 of this manual.

7 Assembly

Note: Ensure that all parts are clean and dry before assembly.
1 Fit the spindle to the body. Fit the lever arm, new tab washer and spindle nut. Check that when the stop on the lever arm is against the abutment on the body, the countersunk ends of the holes in the spindle face toward the intake end of the carburettor. Tighten the spindle nut and lock with one of the two pointed tabs. This enables the tab washer to be re-used. Bend the square tab over the lever arm.
2 Slide the throttle disc into the slot in the spindle in the same relative position marked on disassembly, and fit two new throttle screws. Do not tighten at this stage. Close the throttle when the disc will centre itself in the bore. Check visually that contact is made between the disc and the bore throughout its circumference. Check with the throttle closed that there is clearance between the throttle lever and the carburettor body. Without moving the disc in the slot, tighten the screws and spread the split ends sufficiently to prevent the screws coming undone.
3 Screw the needle valve seating into the float chamber lid. Do not overtighten. Insert the needle into the seating and check that the spring-loaded plunger in the needle operates freely. Fit the float assembly and retain with hinge pin. Check that the lever hinges freely. With the float assembly lever resting on the needle valve check that the gap between the float lever and the rim of the float chamber lid as shown in Fig. HS3. The gap can be adjusted by bending the lever.
4 Fit a new float chamber lid gasket (do not use jointing compound) and fit the lid to float chamber in the same relative position marked on disassembly. Fit and evenly tighten securing screws. Fit identity tag under one of these screws.
5 Fit the float chamber assembly to the carburettor body with spacers between the two as necessary. Ensure that the registers on the body and the chamber engage correctly. If the chamber is flexibly mounted, ie with rubber spacer(s), use new items. Fit and tighten the retaining bolt but take care not to overtighten.
6 Fit the piston lifting pin, spring, new rubber sealing washer (if applicable) and circlip to body.
7 Fit the needle to the piston. The relative position of the two parts is critical and may be either of two arrangements according to the contour of the needle at junction of the taper and the shank. See Fig. H5.

Fit and tighten a new locking screw. Check the security of the piston key in the body. Fit the piston assembly to the body taking care not to damage the needle. Fit the piston spring to the piston rod. Lightly oil (SAE 20W/50) the outside of the piston rod and fit the suction chamber in the same relative position as marked on disassembly. Fit and evenly tighten the retaining screws.
8 Fit the jet bearing, washer and locking nut. Do not tighten the nut. Insert the jet into the bearing and the fuel feed pipe connection (without gland and washer) into the outlet from the float chamber. With the piston assembly in its lowest position and the jet lifted fully, tighten the jet locking nut.
9 Remove the jet. Fit the spring and jet adjusting nut to the jet bearing. Fit the gland, washer and ferrule to the flexible pipe. Check that the end of the tube projects at least 0.188 in/4.8 mm beyond the gland. Lightly smear petroleum jelly (Vaseline) on the outside of the jet and insert into bearing. Insert the feed pipe into the float chamber outlet and tighten the sleeve nut until the neoprene gland is compressed (when the pipe will be held

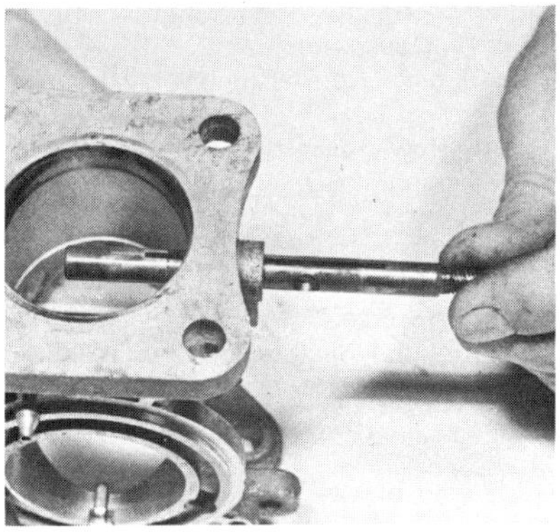

7.1 A little oil helps the spindle

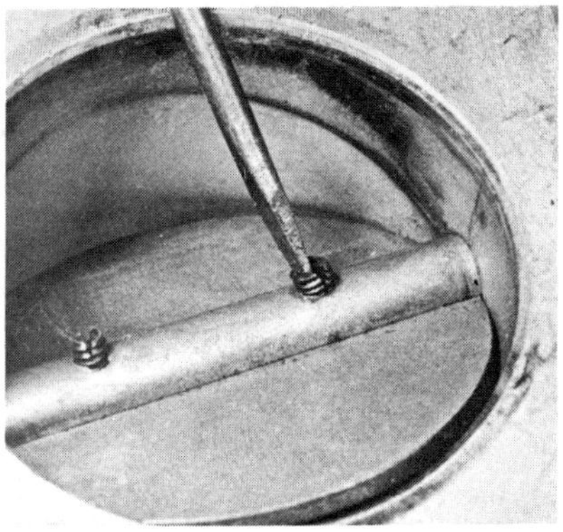

7.2 Use new screws; second hand ones will break

7.3 Here the float is installed before measurement in the traditional way

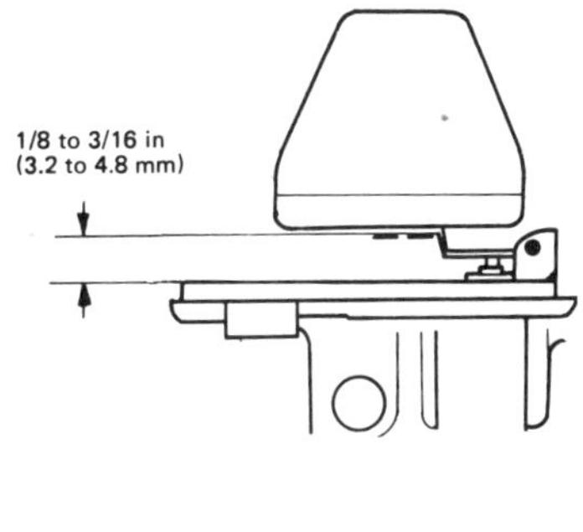

Fig. HS3. Setting float level

the gap should be between 0.125 and 0.1775 in (3.2 and 4.8 mm)

7.5 Clean the mating faces carefully before assembly

7.7 A close-up of the piston locking screw

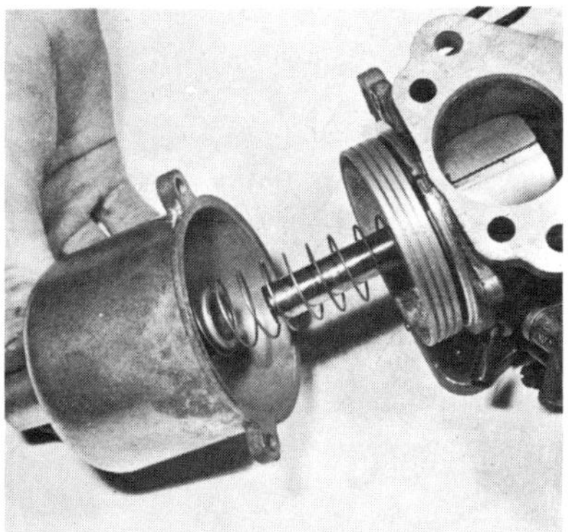

7.7a Clean, slow and careful assembly add to long carburettor life

7.8 Start the thread by hand, then use a spanner

7.9 One of the most delicate threads on the SU carburettor

7.9a Follow the jet centring instructions explicity

firmly in the outlet). Do not overtighten as this can cause leakage and possibly stripped threads in the float chamber. Fit the damper and washer but do not fill piston rod with oil at this stage.

10 Assemble the pick-up lever, cam lever, cam lever spring, skid washer and pivot bolt tube(s). Place the lever return spring in position over the boss and fit the linkage assembly to the carburettor body with the pivot bolt. Ensure that the double-coil spring washer or spacer (alternative part) fits over the projecting end of the pivot bolt tube. Register the angled end of the lever return spring in the groove in the pick-up lever, and hook the other end of the spring around the moulded peg on the carburettor body. Fit the brass ferrule to the hole in the end of the pick-up lever. Relieve the torsion of the return spring and fit the link to the jet with its retaining screw. Fit the baffle plate to the float chamber overflow outlet pipe.

8 Tuning

Note: Refer to Part 1, Chapter 4 of this manual for notes on tuning, with particular reference to preliminary procedures.

Single carburettor installations

1 Set throttle adjusting screw.
Warm engine up to working temperature. Switch off engine. Unscrew the throttle adjusting screw until it just clears its stop and the throttle is shut. Set throttle adjusting screw one-and-one half turns open. Ensure that fast-idle screw is clear of fast-idle cam.

2 Centre jet.
Mark for reassembly and remove suction chamber and piston. Disconnect mixture control wire. Screw up the jet adjusting nut until the jet is flush with the bridge of the carburettor (or fully up if this position cannot be attained). Refit piston and suction chamber. Check that the piston falls freely onto the bridge when the lifting pin is released.

3 Set jet adjusting nut.
Screw down jet adjusting nut two turns (twelve 'flats' on the nut).

4 Set idling speed and mixture strength.
Start the engine and adjust throttle adjusting screw to give desired idling. Turn the jet adjusting nut up to weaken or down to richen until the fastest idling speed consistent with even running is obtained. Adjust again the throttle adjusting screw to give correct idling, if necessary.

5 Check mixture strength. (See Fig. H11 and H12).
The effect on exhaust smoke of mixture strength.

(w)	Too weak	Colourless, irregular note, splashy misfire.
(c)	Correct	Regular and even note.
(r)	Too rich	Blackish, regular or rhythmical misfire.

Check for correct mixture by slowly pushing the lifting pin up about 0.031 in/0.8 mm. The graph illustrates the effect on engine rpm and indicated mixture strength as the piston is raised.

(r)	Rich mixture	rpm increase considerably.
(c)	Correct mixture	rpm increase very slightly.
(w)	Weak mixture	rpm immediately decrease.

6 Connect and set mixture control wire.
Connect the mixture control wire with about 0.0625 in/1.6 mm free movement before it starts to pull on the jet lever. Pull the choke knob at the facia until the linkage is about to move the carburettor jet operating arm and adjust the fast-idle screw to give an engine speed of about 1000 rpm when the engine is hot. Return the choke knob and check that there is some clearance between the fast-idle screw and the throttle stop.

7 Fill damper.
Finally top up the piston damper with thin engine oil grade SAE 20W/50 until the level is 0.5 in/13 mm below the top of the hollow piston rod. *Note:* On non-dustproofed carburettors, identified by a vent hole in the piston damper top, the oil level should be 0.5 in/13 mm above the top of the hollow piston rod.

Multiple carburettor installations

See this section for the Type H carburettor.

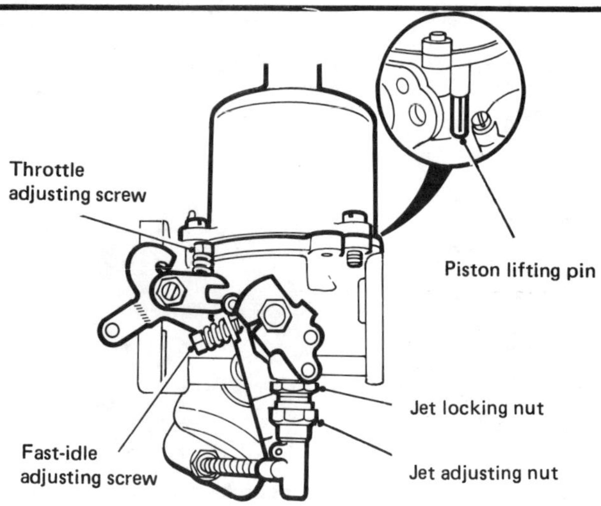

Fig. HS4. Type HS adjustment and tuning facilities

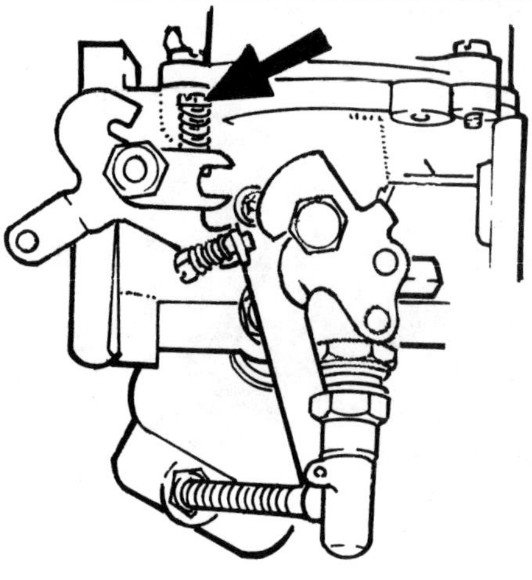

Fig. HS5. The throttle adjusting screw

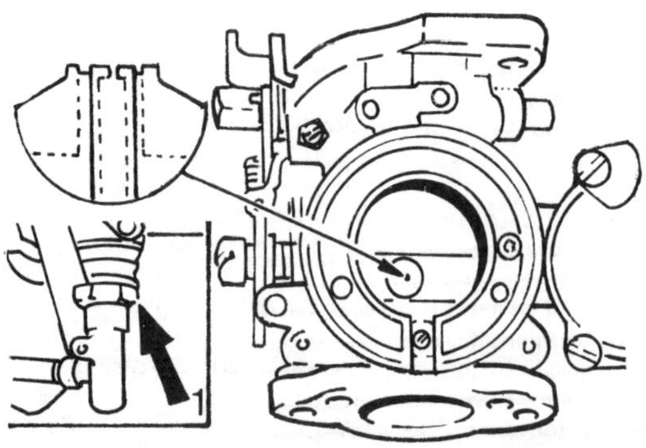

Fig. HS6. Jet centring *1 jet adjusting nut*

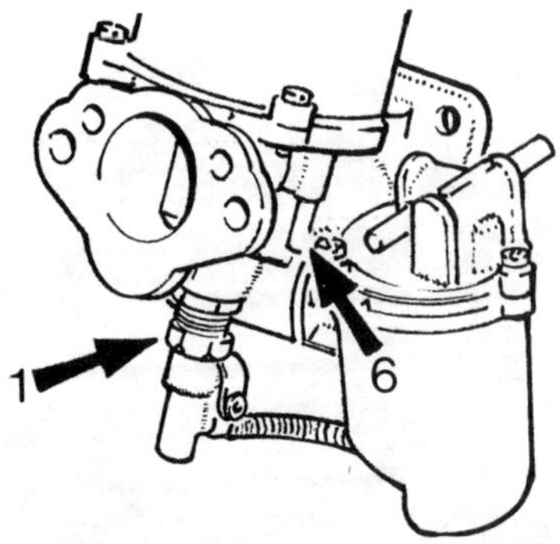

Fig. HS7. Jet adjusting nut setting

1 jet adjusting nut 6 piston lifting pin

Type HIF

Contents

1 Introduction

The Type HIF (Horizontal Integral Float Chamber) carburettor, sometimes known as the 'swinging needle' carburettor, is the most recent development of the SU instrument working on the constant-vacuum principle. It has been designed primarily to meet the requirements of exhaust emission control carburation systems. Functionally similar to previous types, the instrument uses the variable choke/constant depression principle to achieve the precise mixture ratio required to control the toxic elements of exhaust emission to within statutory requirements. It differs from previous types in that additional mixture required for cold starting is supplied separately to that from the main jet. It is manufactured in the larger sizes only; HIF4, HIF6 and HIF7 being 1.5 in, 1.75 in and 1.875 in throttle bore diameters, respectively.

The other characteristic feature of the unit is the absence of a separate float chamber. The float and inlet valve mechanism are contained in a housing underneath the carburettor body, giving the unit a distinctive appearance. This housing also contains a device which alters the main jet position in relation to the metering needle (thereby automatically adjusting the mixture ratio) to compensate for changes in fuel viscosity due to changes in fuel temperature (see Part 1, Chapter 1). This enables the carburettor to maintain a very accurate mixture ratio control over a range of operating conditions.

An over-run valve is incorporated in the throttle disc to limit the depression when the throttle is closed.

2 Float chamber assembly

The float chamber, incorporated in the body casting below the choke bore, is sealed by a removable cover-plate and rubber gasket. It houses a moulded plastic float hinged upon a pivot screwed into the wall of the housing, a spring-loaded needle valve assembly, and the jet operating mechanism.

3 Mixture control

The mechanism consists of a right-angled adjusting lever riveted to a bi-metal blade, the end of which engages with the base of the jet assembly. The lever is flexibly secured to the body by a spring-loaded screw and is adjusted by the jet adjusting screw in contact with one of its limbs. Screwing in and out (from outside) of the jet adjusting screw lowers and raises the jet respectively, giving a fine degree of mixture control. At any position of this lever the bi-metal assembly has an over-riding control function and will compensate for variation in fuel viscosity due to temperature changes.

Note that the height of the jet is not adjusted for cold-start conditions, this function being performed by a separate device. Once the jet height has been set, no further adjustment is necessary and to frustrate attempted tampering, provision is made for fitting a sealing plug in the adjusting screw tapped hole.

4 Jet assembly

The jet assembly differs from previous types chiefly in respect of its redundancy in cold-start conditions. It also differs considerably in construction, being of aluminium alloy and not brass, and has a separate pressed-in brass orifice forming the jet proper.

The integral plastic moulding at its lower end forms an inlet for fuel, an articulated connection for the end of the bi-metal jet lever (which controls its height), a stop to limit its upward movement, and a means of identification (of right or left-hand interconnection, see figure above). The size of the jet (0.090 in or 0.10 in) is given by the absence or presence of a machined groove at the top of the jet tube. The jet assembly is not repairable and must be replaced by a new part complete if damaged.

5 Cold-start enrichment device

Additional mixture required for cold-starting is supplied by a separate cold-start valve and is independent of the main jet.

The cold-start valve consists of a starter valve body, a valve spindle which rotates through a limited arc within it, and an O-ring and V-seal to seal the valve body in its housing, and to seal the valve spindle in the valve body, respectively. A metal seal cover is fitted to prevent damage to the V-seal. The assembly is fitted into a bored-out housing in the side of the carburettor and is operated by a lever and a return spring.

The valve body has a hole drilled through its wall which communicates, via the annular space in the housing bore, with a fuel supply passage. An air bleed hole breaks into this passage above the fuel level (controlled by float).

The spindle is hollow and communicates with the passage which terminates in the carburettor mixing chamber at the back of the bridge. A hole in the wall of the spindle corresponds with that in the valve body when the spindle is in a certain position. At each side of the hole is a tapering, machined groove. The varying depth of this groove gives a progressive throttling effect as the spindle is turned to different positions.

6 Piston and needle

The spring-loaded needle assembly is secured in the piston by the needle locking screw. A flanged collar at the top of the

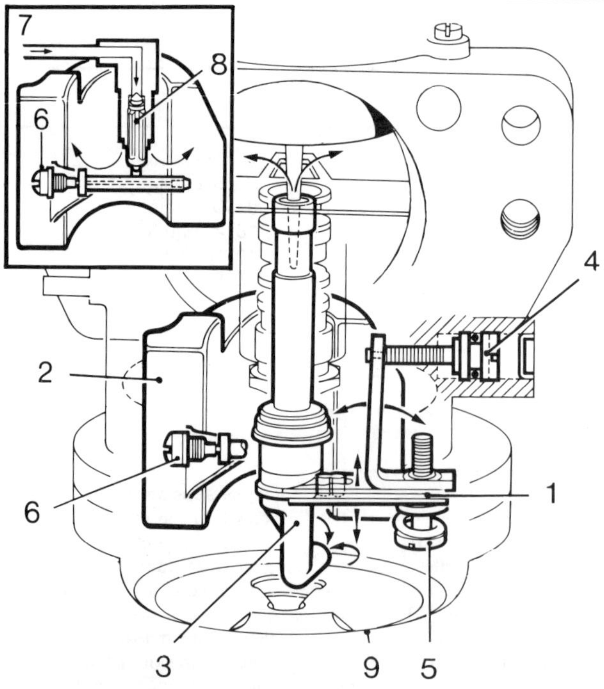

Fig. HIF2. The float chamber layout, unique to the Type HIF

1 bi-metal assembly
2 concentric float
3 jet head
4 jet adjusting screw
5 bi-metal pivot screw
6 float fulcrum screw
7 fuel inlet
8 needle valve
9 bottom cover-plate

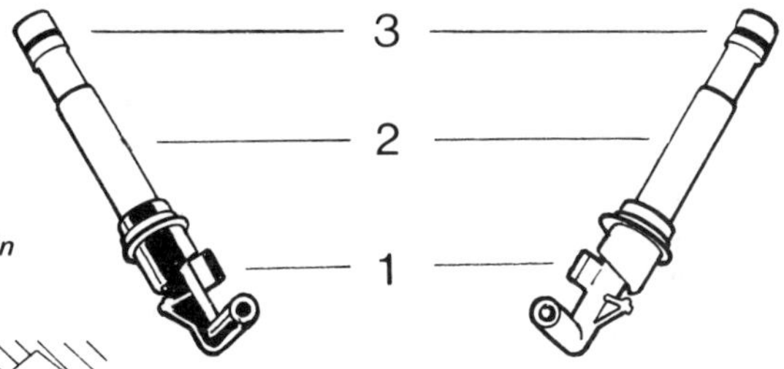

Fig. HIF3. Jet identification

left - right hand interconnected
right - left hand interconnected carburettors
1 jet head (black - right) (white - left)
2 jet assembly
3 without groove - 0.9 in with groove - 0.1 in

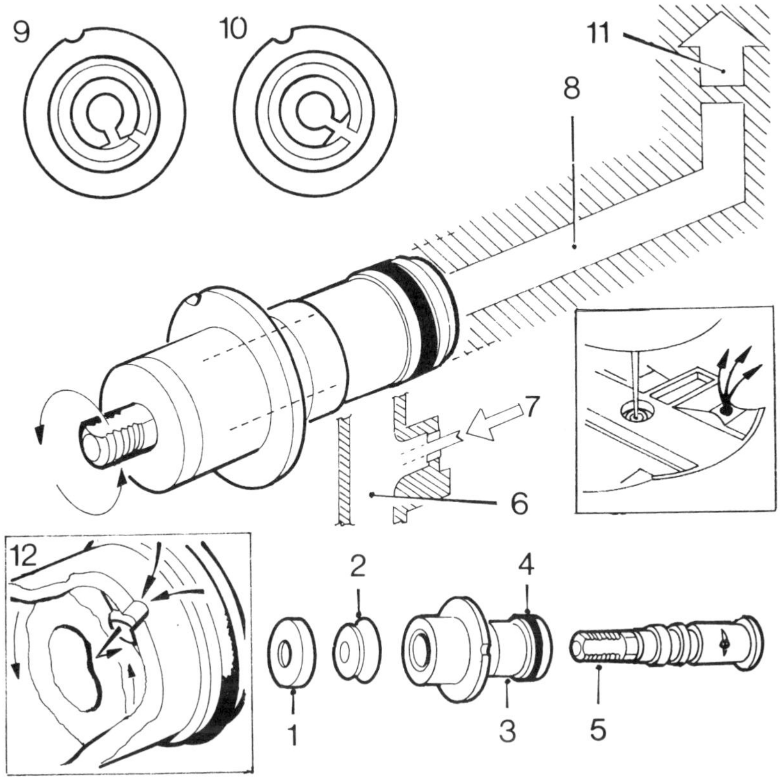

Fig. HIF4. Cold start enrichment device

1 End seal cover
2 End seal
3 Starter valve body
4 'O' ring
5 Valve spindle
6 Fuel supply
7 Air bleed
8 Fuel delivery to jet bridge
9 Commencement of enrichment
10 Maximum enrichment
11 Enrichment outlet
12 Fuel flow through valve

needle bears against a protrusion on the needle guide, which tilts it under the action of the spring. The needle is thus biased towards a particular position in the jet, either forwards or backwards depending upon design of needle guide (location of protrusion). An etched alignment mark on the underside of the needle guide ensures correct assembly. The mark must be between the transfer holes.

7 Operation

Cold-starting

With the choke control (on the car's instrument panel) pulled fully out, the cold-start valve is rotated to its fully open position when the hole in the inner spindle aligns with the hole in the valve body, providing a maximum free area for fuel flow. The fuel level in the feed passage is below the air bleed and there is no flow of fuel until a depression is generated in the mixing chamber of the carburettor.

When the depression is generated by turning over the engine (hand cranking or on starter motor), fuel is drawn up out of the feed passage and mixes with air drawn through the air bleed to form an emulsion in the annulus surrounding the valve body. The mixture flows through the port in the valve body, through the corresponding hole in the inner spindle, through the hollow spindle and is discharged into the mixing chamber of the carburettor.

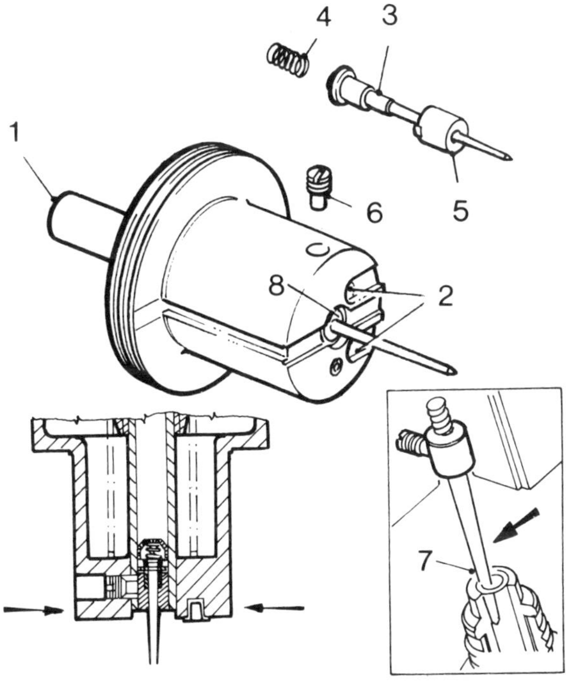

Fig. HIF5. The swinging needle fixing

1 piston rod
2 transfer holes
3 jet needle
4 needle spring
5 needle guide
6 needle locking screw
7 needle biased in jet
8 etch mark

Gradual return of the choke control to its fully home position results in a correspondingly progressive reduction in rich-mixture quantity delivered. As the main air valve will also have opened when the engine fired, the net mixture strength will also weaken.

Full throttle

With the engine at normal running temperature and the cold-start valve out of action, the carburettor will function as a normal H-Type unit, mixture quantity and strength delivered being dependant upon the interaction of the throttle opening, the manifold depression and the fuel needle/air valve relationship.

Acceleration

Temporary enrichment for acceleration is provided by the resistance to upward movement of the air valve, caused by the hydraulic damper. It is single acting and has no effect on deceleration, allowing the air valve to respond immediately, thus preventing an over-weak mixture being delivered with its attendant ill effects on stability of combustion.

Overrun

An overrun condition obtains when the vehicle is 'driving the engine', which will occur, for example, when descending a hill with the throttle closed. In this condition very high manifold depression is produced which is not conducive to efficient, 'clean' combustion. The maximum depression is governed by the operation of a spring-loaded plate valve in the throttle disc which opens at a pre-determined value to admit air into the mixing chamber and thus reduces the depression.

Part throttle condition

At small throttle openings the mixture is conducted through a small bore passageway below the 'floor' of the main choke bore, to emerge at a point adjacent to the bottom edge of the throttle plate, where a cut-out is formed. This results in the mixture velocity being considerably higher than if it were induced normally, due to the much smaller cross-sectional area, with more complete atomization of the mixture, particularly at the local high depression at the throttle plate cut-out. This arrangement gives greatly improved combustion and therefore minimum toxic emissions.

Fuel temperature variation

At all phases of operation described above, the precise mixture ratio is also subject to the height of the main jet, which is controlled by the fuel temperature compensator. With increasing fuel temperature the jet is raised, and with decreasing temperature it is lowered. Thus the variation in fuel flow which would normally ensue as a result of the change of viscosity is compensated for by an inverse and proportional change of annular fuel flow area, and the actual flow is maintained at a constant value.

8 Overhaul

It is strongly recommended that overhaul of this unit is restricted to disassembly, cleaning and re-assembly. Damaged parts should be replaced and no attempt should be made to repair any item which has a regulating or metering function.

The unit, despite having the appearance of a slightly different but conventional type, is incomparably more sophisticated and may not be satisfactorily dealt with by the methods detailed in this manual. It has been specifically developed to meet the requirements of toxic emission regulations and is manufactured to far finer dimensional tolerances than previously employed.

A damaged unit may be restored to a serviceable condition by replacement of defective items, but subsequent tuning demands the use of specialized equipment in order to obtain the performance of which this unit is capable.

Reconditioning of a generally worn unit is not feasible and it should be returned to the manufacturer for overhaul, or for replacement with a new unit.

9 Disassembly

1 Unscrew damper cap (standard suction chamber only) and remove piston damper. Remove and discard damper washer.

Standard suction chambers

2 Remove the screws securing the suction chamber. Collect the identity tag. Holding the unit upright, remove the suction chamber, piston spring and piston. Note that it is not necessary to mark the suction chamber and body for correct relative position as they can only be fitted together one way due to the unequal disposition of lugs.

Ball bearing suction chambers

Remove the screws securing the suction chamber. Collect the identity tag. With the unit upright, hold the piston firmly in the choke. Lift the suction chamber until the damper retainer is freed from the piston rod. Remove the damper and from the suction chamber, remove the piston spring and piston.

3 Remove and discard the needle retaining screw and withdraw the jet needle, needle guide and needle spring. Separate these components, taking care when the spring is removed, that it is not stretched.

4 Remove the circlip from the lower end of lifting pin and collect the lifting pin spring. Discard the circlip. Remove the pin.

5 Remove the cover screws. Remove and discard the spring washers. Remove the float chamber cover. Note that it is not necessary to mark the cover for correct relative position as the incorporation of fouling pegs ensures its correct location on refitting. Remove and discard the float chamber cover seal.

6 Remove the jet retaining screw and jet spring. Withdraw the jet assembly and bi-metal jet lever together. Separate the jet assembly and lever taking care not to damage the plastic moulding at base of the jet assembly.

7 Remove the float pivot. Remove and discard the pivot seal. Remove the float. Invert the carburettor when the float needle will drop out. Remove the jet adjusting screw. Remove and discard the adjusting screw seal. Remove the float needle seat using a spanner.

8 Remove the jet bearing nut and withdraw the jet bearing. Remove and discard the jet bearing washer.

9 Remove and discard the throttle disc screws. Open the throttle and pull the throttle disc from the slot in the throttle spindle. Remove any burrs around the screw holes with a fine file to prevent damage to the PTFE bushes as the spindle is withdrawn.

10 Withdraw the throttle spindle. Remove and discard the throttle spindle seals. Remove the throttle spring. Bend back the tabs of the tab washer and unscrew the retaining nut. Remove and discard the tab washer. Remove the throttle lever and the throttle actuating lever. Remove the throttle adjusting screw and nut and fast-idle screw and nut from the throttle actuating lever.

11 Bend back the tabs of the tab washer and remove the retaining nut. Remove and discard the tab washer. Remove the fast-idle cam, cold-start spring, end cover and spindle seal. Discard the spindle seal. Remove the retaining screws and retaining plate.

12 Remove the cold-start body complete with the cold-start spindle. Remove and discard the O-ring and cold-start seal.

13 Do not remove the piston key (fitted to the periphery of piston bore at top of the casting), or the throttle spindle bushes unless these parts require replacement. Do not remove any brass orifices, plugs or stub pipes.

10 Assembly

Note: Ensure that all parts are clean and dry before assembly.

1 Fit a new O-ring to the cold-start body. Insert the cold-start spindle into the body and fit a new spindle seal over the threaded end of spindle. Fit the seal with the thin, soft flange inwards

9.1 An unassembled damper rod. Discard the washer actually resting on the rod

9.2 The damper has been left assembled; the suction chamber can only be fitted one way

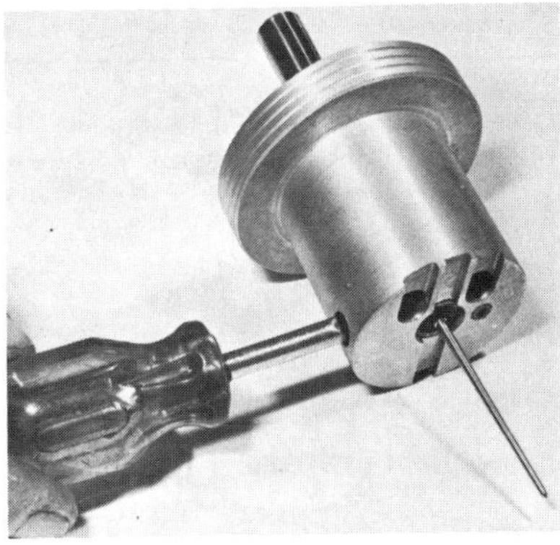

9.3 The traditional procedure is still used

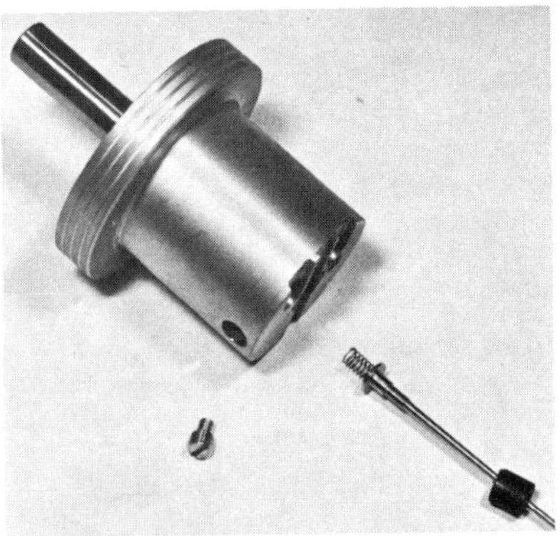

9.3a Note all the components; check the spring carefully

9.5 The float chamber cover only fits one way

9.6 Note the huge jet retaining clamp spring

9.6a This shows well the float around the jet base

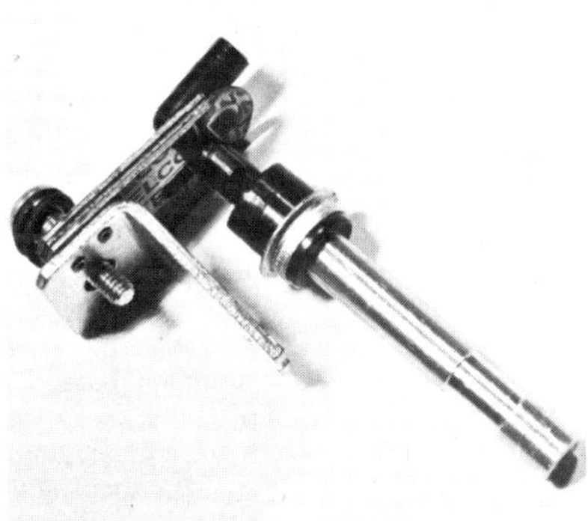

9.6b The complete jet assembly

9.7 The float hinge pin screws in this time, from outside

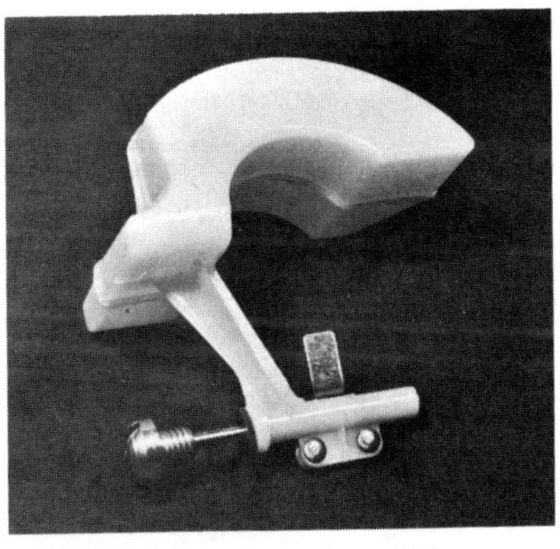

9.7a There is certainly no repair for a punctured or fractured float

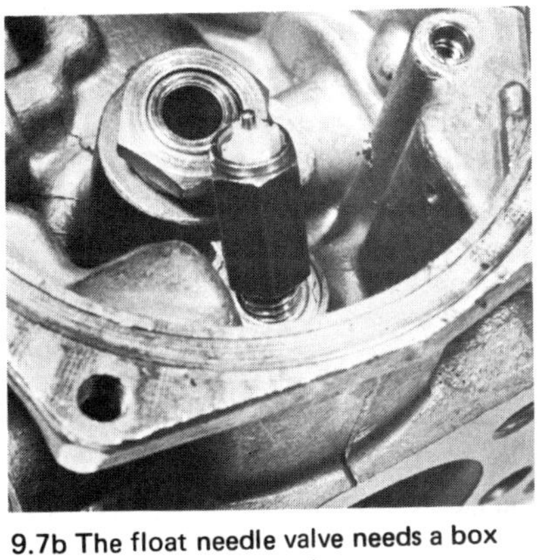

9.7b The float needle valve needs a box spanner for easy removal

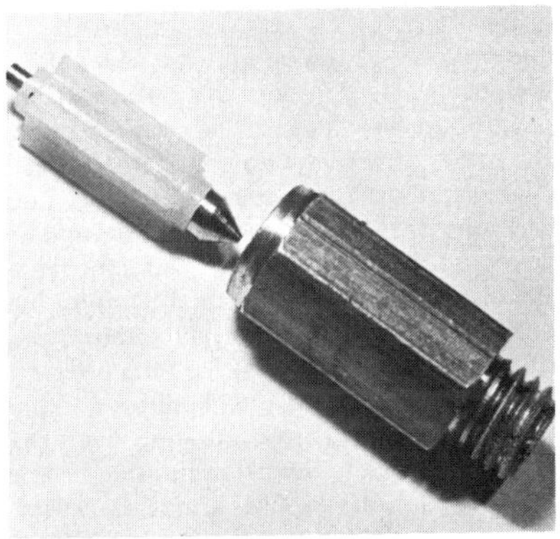

9.7c The two components of the float needle valve

9.8 Never re-use the jet bearing washer

9.8a The jet components

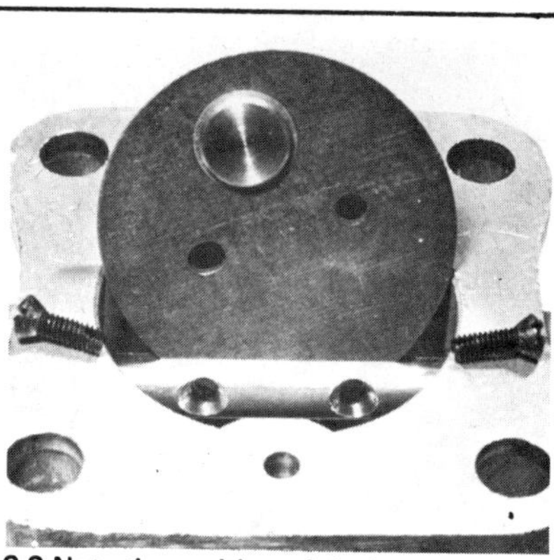

9.9 Note the position of the butterfly here

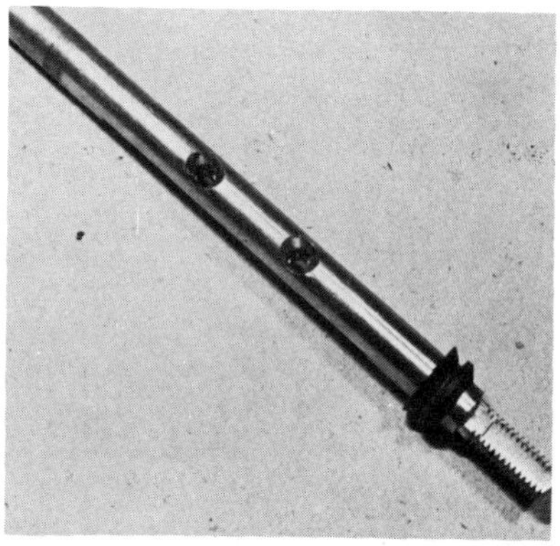

9.10 Always use new throttle spindle seals

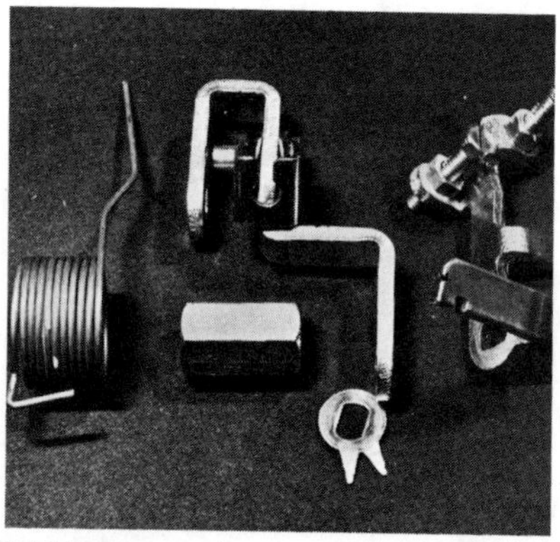

9.11 The external components of the cold start device

9.12 Discard the seal and O-ring; use new on assembly

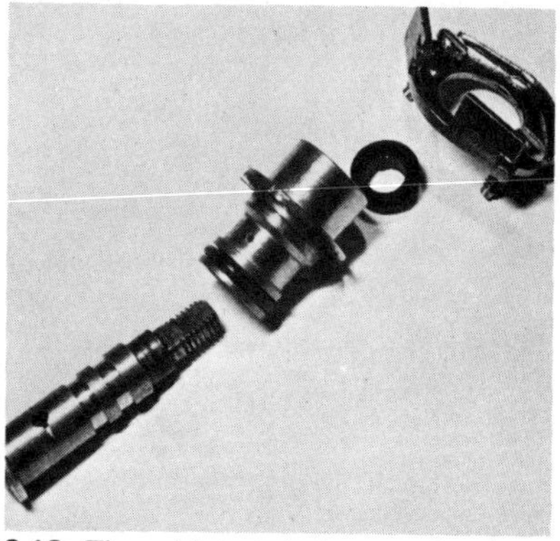

9.12a The cold start device complete in component form

and press up to the body. Fit a new cold-start seal to the body and insert assembly into the carburettor body. Note that the cut-out on the flange of the cold-start body must register with the uppermost screw hole in the carburettor body.

2 Fit the retaining plate (large flange outward and facing carburettor outlet flange) with the retaining screws. Fit the end cover, cold-start spring (with straight tail engaged in the slot in the upper flange of the retaining plate), fast-idle cam, tab washer and retaining nut. Tighten the retaining nut and bend the tabs of the tab washer against the flats on the nut to lock.

3 Fit the throttle adjusting screw and nut and fast-idle screw and nut to the throttle actuating lever. Do not lock the nuts at this stage. Fit the throttle actuating lever and throttle lever to the throttle spindle. Check that the levers are fitted correctly to the spindle (the countersunk ends of the screw holes in the spindle must face away from the levers). Fit the tab washer and retaining nut. Tighten the nut and lock with the tabs. Fit the throttle spindle seal to the throttle spindle (thin, soft flange toward centre of spindle) and fit the spindle to the carburettor body. Fit the throttle spindle seal to the exposed end of the spindle and press up into the recess in the carburettor body. Fit the throttle spring over the retaining nut and allow the straight tail to rest against the underside of the throttle lever. Pre-load the spring by rotating cranked end one turn clockwise and engage the hook in the slot in the throttle lever. Engage the straight tail with the outer slot in the lower flange of the retaining plate.

4 Open the throttle (turn levers through 90° against the spring load) and slide the throttle disc into the slot in the spindle. Note that the throttle disc must be fitted with the head of the over-run valve towards the engine flange, ie above the throttle spindle when the valve is closed. Close the throttle to centre the plate in the bore. Check that the cut-out at the bottom of the disc registers. Fit new throttle disc screws. Adjust the throttle disc until it is accurately centred in the bore and tighten the screws. Spread the ends just sufficiently to prevent them unscrewing.

5 Fit a new jet bearing washer to the jet bearing. Fit the jet bearing to the carburettor body with the jet bearing nut. Tighten the nut.

6 Fit the float needle seat using a spanner. Do not overtighten. Fit a new adjusting screw seal to the jet adjusting screw and screw into the carburettor body until the end of the screw is flush with the inside wall of the float chamber. Fit the float needle into the float needle seat. Fit a new pivot seal to the float pivot. Fit the float and secure with the pivot seal.

7 Check the float level by inverting the carburettor and checking that the point indicated on the float is 0.04 in (+ 0.02 in) 1.0 mm (+ 0.5 mm) below the face of the float chamber. Bend the brass tab if necessary to adjust the setting (See Fig. HIF6).

8 Assemble the jet assembly and bi-metal jet lever. Insert the jet assembly into the jet bearing and secure the bi-metal jet lever to the carburettor body with the jet spring and jet retaining screw.

10.1 Assemble as shown here; all must align properly

10.1a Note how the spring fixes into the slots

10.2 The first view of the complete cold start device installed ...

10.2a ... the second view finally explains its location

10.2b As a treble check to ensure accuracy of assembly

10.3 Lightly oil the spindle before assembly; use new seals

10.4 The other side of the butterfly almost re-assembled

10.5 Note the new seal on the end of the jet base

10.6 The needle valve is finally installed; tighten with a box spanner

10.6a The float is installed and the hinge pin tightened

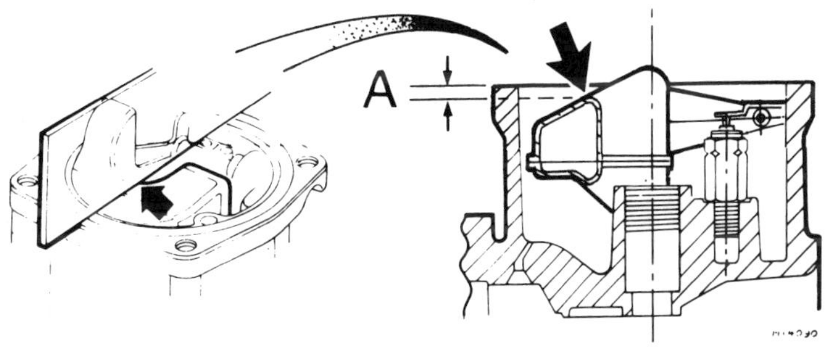

Fig. HIF6. Float level checking dimension (A); checking point indicated by arrows

9 Fit a new float chamber cover seal and the float chamber cover with new spring washers and cover screws. Tighten the screws evenly, in a diagonal sequence.
10 Fit the needle spring to jet needle and assemble to the needle guide. Note that the pip on the needle guide must be in contact with the flange on the jet needle and that the scribed line on the lower face of the needle guide must be parallel with, and between, the milled channels at the bottom of the piston. Fit the assembly to the piston with a new needle locking screw.
11 Fit the piston to the carburettor body. Fit the piston spring and suction chamber. Fit the identity tag and screws. Fit a new damper washer to the damper cap and screw in the piston damper.

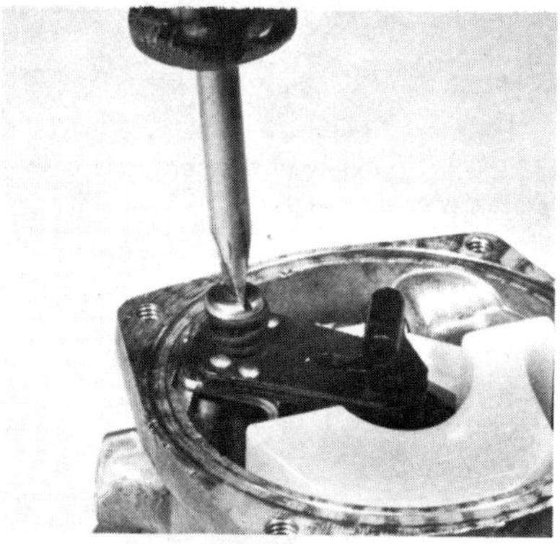

10.8 Final jet assembly means careful, tightening of the jet retaining screw

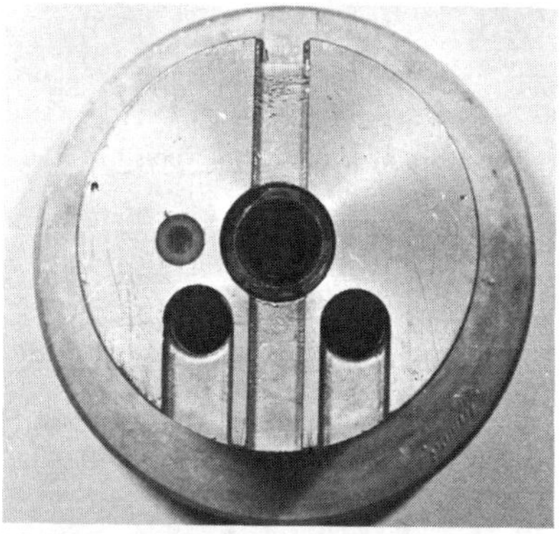

10.10 The bottom of the piston; make sure all is very clean before fixing the needle

10.11 The piston is now properly located

11 Tuning

Due to the complexity of this instrument it is strongly recommended that no attempts are made at tuning by the amateur. This does not imply that the amateur has insufficient knowledge or skill to perform the task; but rather has insufficient test facilities and equipment.

It will be fully understood that in order to realise the performance of which this type is capable, the tuner must have a means of analysing the exhaust gas, and of accurately measuring temperature (ambient and fuel), pressure and air flow. The tuning and balancing of multiple carburettor installations introduce further complications.

However, this advice assumes that it is required to tune the carburettor to perform the duty for which it was designed, ie to provide a carburation system which gives minimum toxic emissions. If the carburettor is merely required to provide normal carburation, then it may be tuned in the usual way and the refinements ignored, since at worst they will have no deleterious effects on performance, and at best may (more by luck than judgement) yield some of their benefits and give something approaching correct mixture ratios.

12 Tuning procedures

Single carburettor installations

1 Unscrew the throttle adjusting screw until it is just clear of the throttle lever with the throttle closed. Check that the cold-start lever is fully closed (cold-start device out of action) and that the fast idle adjusting screw is well clear of the cam. Failure to observe this may result in the throttle being held open when, by external examination, it appears closed. Turn the throttle adjusting screw one and a half turns clockwise to open the throttle for a datum position.
2 Lift the piston with the piston lifting pin, checking that it falls freely when the pin is released.
3 Lift the piston high enough to reveal the bridge and jet. Turn the jet adjusting screw to bring the top of the jet flush with the bridge, or as near the top as possible.

Turn the jet adjusting screw two turns clockwise to set the jet height to a datum position. Repeat the test detailed in paragraph 2 above, to check for freedom of movement of the piston assembly.
4 Check the piston damper oil level as follows:
a *Standard suction chambers;* Remove the damper and fill the piston rod with SAE 20 oil to 0.5 in/13 mm above the top of the hollow piston rod. Refit the damper. Do not overtighten.
b *Ball bearing suction chambers;* Unscrew the damper cap and raise the piston and damper to the top of their travel. Fill the damper retainer with SAE 20 oil and lower the damper until the cap contacts the suction chamber. Repeat this procedure until the oil level is just visible at the bottom of the retainer recess. Refit the damper. Do not overtighten.
5 Start and run the engine until it reaches normal working temperature, and continue to run for approximately five minutes, in order to ensure that the engine is 'soaked' at the correct temperature and that an even temperature gradient is established.

Increase the engine speed to 2,500 rpm for approximately 30 seconds. This will clear the plugs and clear any condensed fuel from the inlet manifold walls.
6 Set idle speed.

Set the correct idling speed (as given by engine manufacturer) by adjustment of throttle adjusting screw.
7 Set mixture strength.

Note that on some units no provision is made for adjusting the mixture strength as the adjusting screw is sealed by a plug (see description).

Set the correct mixture, strength by adjustment of the jet adjusting screw (turn clockwise to enrich and counter-clockwise to weaken) until the fastest engine speed (with given throttle adjusting screw position) is obtained. This fastest engine speed must be obtained when adjusting from a weak setting and approaching a rich setting, so having found the correct setting, turn the screw counter-clockwise until a small but perceptible fall in speed is noticed, then turn the screw clockwise very slowly until the maximum speed is regained. If the correct position is 'overshot', open the throttle to clear the engine, return to a wak setting and repeat the tuning procedure.
8 Re-adjust idle speed.

Following the mixture strength setting, the idle speed may need re-adjusting by adjustment of the throttle adjustment screw to restore the correct speed.
Note: It is at this stage that the exhaust gas analyzing test

would be applied, if the tuning were being performed by an operator with this facility. The mixture strength obtained by very careful application of the procedures given above gives only an approximation to correct carburation; final adjustment being made in conjunction with the exhaust analysis equipment to bring the carbon monoxide (CO) percentage within the required limits. If this entails more than half a turn of the adjusting screw, the carburettor is considered unserviceable. Such is the accuracy of carburation of which this instrument is capable, and which is demanded by certain authorities.

9 Set fast idle.

With the cold-start device fully closed (fast idle cam against its stop) check that there is 0.060 in/1.5 mm free movement of the mixture control (choke) cable before the cable moves the cam.

Pull out the mixture control (choke) on the vehicle instrument panel until the arrow marked on the cam is directly under the fast idle adjusting screw. Turn the screw to obtain the correct fast idle speed.

10 Refit air cleaner/intake system.

Lock both adjusting screws with locknuts, taking care not to disturb the settings.

Multiple carburettor installations

1 Slacken throttle spindle and cold-start valve interconnections.

2 Follow instructions given in paragraphs 1 to 5 for single carburettor installations.

Note that it is vital that any adjustments made to the carburettors are equal on both (all).

3 Set idle speed.

Set the correct idling speed (as given by engine manufacturer) by adjustment of throttle adjusting screws. Ensure that the air flow through each carburettor is the same by using a small bore tube, one end positioned approximately 0.25in/6 mm inside the intake and the other end held to the ear, and listening to the intake 'hiss'. A proprietory vacuum meter may be used to achieve this important balancing of the units.

4 Set mixture strength.

Set mixture strength according to the instructions given in paragraph 7 for single carburettor installations. Again ensure that both (all) screws are turned by an equal amount.

5 Re-adjust idle speed.

Re-adjust idle speed according to the instructions given in paragraph 8 for single carburettor installations. Check the balance of the airflows by the method described above, as this is more direct, and therefore more accurate, than observing equal turning of the adjusting screws.

6 Reconnect the throttle spindle interconnecting linkage ensuring that there is clearance between the drive pins on the linkage ends and the lower edges of the forks. Tighten the clamping bolts and check that the rod has approximately 0.050 in/1.3 mm end float.

7 Run the engine at a fast idle and check the balance of the carburettor airflows by the method given in paragraph 3 above.

8 With the cold-start valves fully closed (fast idle cams against their stops) reconnect the interconnecting linkage. Operate control and check that both (all) valves start to move simultaneously. Check that there is 0.060 in/1.5 mm free movement of the mixture control (choke) cable before the cable moves the cams.

9 Set fast idle.

Pull out the mixture control (choke) on the vehicle instrument panel until the arrows marked on the cams are directly under the fast idle adjusting screws. Turn the fast idle adjusting screws equal amounts to obtain the correct fast idle speed.

10 Refit air cleaner/intake system.

Lock all adjusting screws with locknuts, taking care not to disturb the settings.

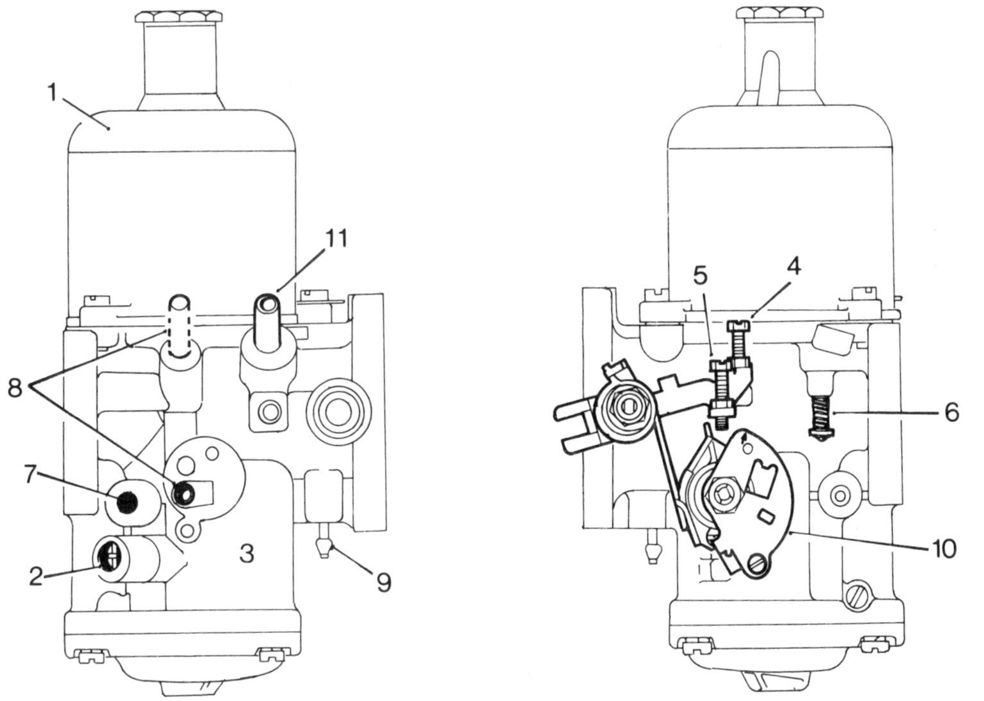

Fig. HIF7. Type HIF adjustment and tuning facilities

1 piston/suction chamber
2 jet adjusting screw
3 float chamber
4 throttle adjusting screw
5 Fast-idle adjusting screw
6 piston lifting pin
7 fuel inlet
8 vent tube (alternative positions)
9 auto ignition connection
10 cold start enrichment lever (cam lever)
11 crankcase ventilation tube

Thermostatic starting carburettor

Contents

1 Introduction

The thermostatic starting carburettor, incorporated in some installations, provides automatically differing degrees of mixture enrichment at starting, idling, cruising and full throttle conditions, when the engine has not attained normal running temperature.

2 Construction

The unit consists of a die-cast aluminium body, a solenoid assembly, a disc valve and a needle valve.

The body has a machined recess to accommodate the solenoid which is secured to it by a spring clip hinged upon lugs, and a retaining screw.

Passages are drilled for air and mixture transfer through the two valves, and screwed union connections are provided for fuel inlet and mixture outlet.

The solenoid assembly consists of a solenoid to which is attached a cover with terminals for connection of 12V dc electrical supply. There is a brass tube in the centre in which slides an armature. The lower end of the armature is provided with a socket which accommodates the ball extension of a brass disc. A light, conical, compression spring, fitted between the disc and a brass spring retainer underneath the solenoid, loads the armature and disc assembly out of the solenoid and seats the valve disc on a circular 'knife-edge' brass seating pressed into the valve body. The articulated ball and socket joint compensates for slight manufacturing inaccuracies by allowing the valve disc to tilt in any plane to ensure full contact on the seating.

The needle valve assembly consists of a brass rod, tapered at its lower end, and fitted with a plunger near the top. It is supported in a screwed bearing (adjustable stop) fitted to a bar which is attached to the top of the body. The plunger acts as a spring register and suction disc. The assembly is fitted into a bore in the body, the tapered end of the needle entering a jet screwed into the bottom of the bore from underneath. A light, helical, compression spring loads the needle out of the jet.

The unit is mounted to the base of the main carburettor by an arm which is centrally drilled for the passage of fuel. The arm has a boss at each end and is secured by banjo bolts. The arm is angled slightly from the horizontal and is reversible (by inverting it) to give varying petrol levels. The outlet of the unit is connected by an external pipe to the inlet manifold.

3 Installation and control

Energization of the solenoid is normally under the control of the ignition switch, and a thermal switch which is usually mounted in the engine coolant pipe to the top of the radiator, or fitted direct to a tapping in the cylinder head.

The switches are wired in series, therefore the electrical circuit is completed when both of these switches are closed (ie when the ignition is switched ON and the engine coolant is cold) and is broken when either of them is open (ie ignition switched OFF or engine coolant hot). The circuit is not normally fuse-protected.

Due to a fault in the thermal switch the unit occasionally remains in operation unnecessarily, and cars are often to be found re-wired so that the unit is under the control of a facia-mounted, driver-operated switch. In this case a warning lamp should also be fitted in parallel with the carburettor as a precaution against accidental operation of the switch, or inadvertently leaving it switched on.

4 Operation

With the engine stationary and the solenoid coil energized the armature is retracted into the bore of the coil against the loading of the conical spring, thus lifting the disc valve from its seat.

When the engine is rotated by the starter, air is drawn through the air intake and past the jet, where it is mixed with fuel. The mixture flows upwards, past the shank of the needle and out through the disc valve into the inlet manifold via the external feed pipe. The level of fuel at the jet is controlled by the main carburettor float chamber inlet valve and is above the jet when the engine is stationary, giving a reservoir of fuel in the well of the auxiliary carburettor. When starting with the device in operation, this fuel is drawn into the induction manifold creating the rich mixture necessary for instant starting.

With the engine running, the needle responds to variation in inlet manifold depression against the load of the compression

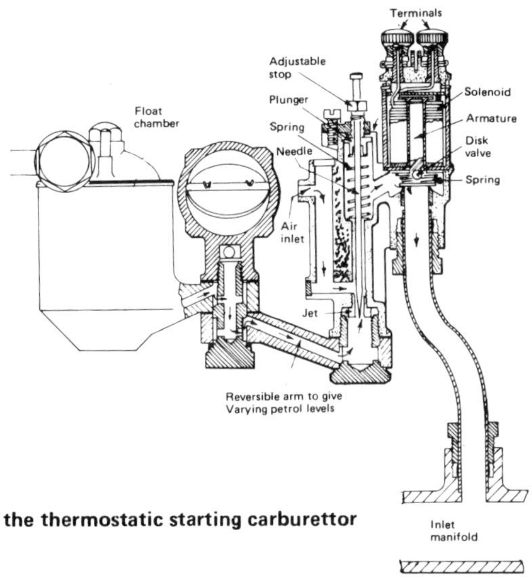

Fig. T2. Cross section through the thermostatic starting carburettor

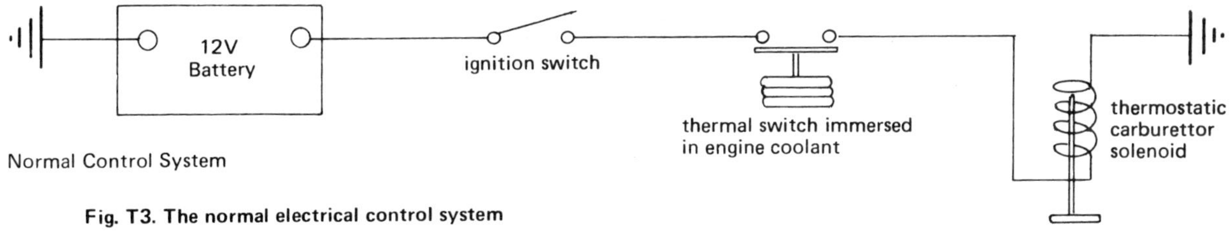

Fig. T3. The normal electrical control system

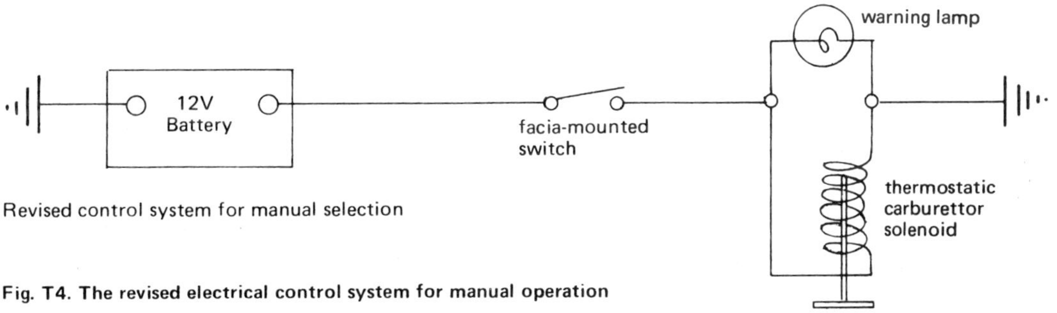

Fig. T4. The revised electrical control system for manual operation

spring. With increasing depression the needle moves downward into the jet until the head abuts the adjustable stop. This weakens the mixture. The mixture will be weakest at high depression (idling with throttle nearly closed) and richest at low depression (large throttle openings). The engine is thus supplied with an additional quantity of mixture at a strength proportional to its demands.

Note that if the solenoid is energized while the engine is idling the disc valve will not normally lift, as the force due to the high manifold depression, combined with the spring load, is greater than the attraction force generated by the solenoid. Momentary opening of the throttle will reduce the depression and permit the valve to lift. It will remain lifted when the throttle is closed.

5 Disassembly

Warning: *Disconnect cables from both terminals of the battery before removing wires from the terminals of the thermostatic carburettor, prior to removal of the unit. When refitting, connect wires to the carburettor terminals before reconnecting the battery terminals.*

1 Disconnect the electrical wiring and remove the feed pipe to the inlet manifold.

2 Unscrew the banjo bolt securing the unit to the mounting arm, and remove the unit from the engine. Collect petrol that drips from the arm.

Do not turn engine over by hand cranking, or otherwise operate fuel pump while unit is removed.

3 Remove the clamping screw from the top of the clip.

4 Hinge the clip to one side, remove the cap and solenoid taking care that the armature and disc valve assembly does not fall out.

5 Remove the armature assembly from the coil, remove the conical spring and spring register.

6 Remove the screws and shakeproof washers. Withdraw the needle valve assembly and spring from the body. Remove the baffle plate.

7 Remove the jet.

6 Inspection

1 Body

Check for corrosion and clean as necessary. Check condition of all tapped holes and repair, using repair schemes suggested for the main carburettor. Check the security of the brass seat insert. Check the security of the plugs at the bottom of the air inlets/fuel reservoirs.

2 Springs

Check the springs for bowing and flattening of coils. Replace a damaged spring. Five different springs are available, identified by a coloured end coil, as follows:

Paint colour on end coil	Load at length (oz)	(in)	Part No
White	*1¾*	*1*	*AUC 1195*
Blue	*2¼*	*1*	*AUC 1041*
Yellow	*2¾*	*1*	*AUC 5021*
Red	*3¼*	*1*	*AUC 3427*
Green	*3¾*	*1*	*AUC 3127*

3 Needle valve

Check that the needle is straight and that the plunger is secure.

4 Solenoid

Check for operation at assembly.

5 Armature/disc valve assembly

Check for corrosion and clean up with fine emery cloth. Check that disc valve articulates freely in armature with no tendency to stick, and that seating face is not grooved. Do not attempt to separate disc from the armature. Check that the armature slides freely in the bore of the coil.

7 Assembly

1 Fit the jet using a spanner.

2 Fit the spring to the needle valve assembly and insert into the body. Fit the baffle plate and secure with screws and new shakeproof washers.

3 Fit the spring register, conical spring and armature to the solenoid.

4 Fit the solenoid to the body.

5 Fit the terminal cover, spring clip and clamping screw. Do not overtighten screw.

6 Refit to the engine, connect wiring and mixture feed pipe. Reconnect the battery cables to the terminals.

7 Switch the ignition ON (engine cold) and OFF checking audibly that solenoid operates at each selection.

8 Tuning

Tuning of the thermostatic carburettor is limited to adjusting the maximum travel of the needle into the jet.

1 Tune the main carburettor, referring to the appropriate section of this manual.

2 With the engine cold, switch ignition ON, check audibly that the auxilliary carburettor operates, and start the engine.

3 Adjust the stop nut by turning clockwise (lowering the needle into the jet and weakening the mixture) until the engine begins to run erratically.

4 Turn the adjusting nut counter-clockwise (lifting the needle out of jet and enriching the mixture) until engine speed has risen to a maximum and fallen to 800 - 1000 rpm, with exhaust gases noticeably black. This is the correct position of the adjustable stop.

5 Check that the thermal switch opens and de-energizes the solenoid when the engine has attained normal working temperature.

6.1 Really check the condition of the thermo body; note the damaged thread here

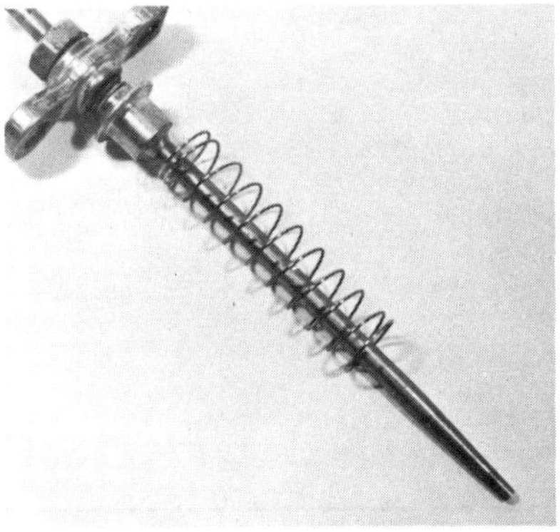

6.3 Check the straightness of the needle, the plunger and the 'spring' of the spring

6.4 The solenoid should be spotless, cleaner than here

6.5 The armature should be this clean at all times

7.1 You must use a spanner; there is no other way

7.2 Use new shake proof washers on each reassembly

7.4 Really clean the base of the armature

7.4a Then reassemble carefully

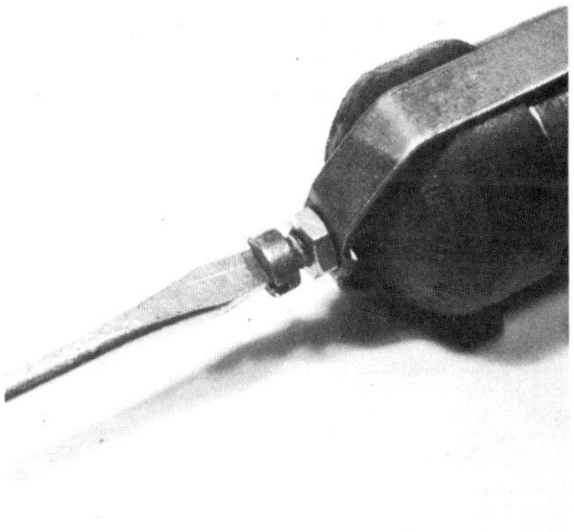

7.5 Do not overtighten this screw: then lock it

7.5a Finally the thermo is ready for assembly to the carburettor

PART 3
Appendix 1 - Applications list

This list gives details of SU carburettors fitted to passenger vehicles of British manufacture from circa 1950 to date. The list is arranged alphabetically in sections according to name of manufacturer. Each section is sub-divided into groups according to models and/or date of manufacture.

The list is presented in tabular format with the following headings:

Model details	Name given to vehicle by manufacturer. May also include a qualifying description eg mark, capacity, export details, manufacturer of special tuning equipment, engine details.
Capacity	Engine capacity in cubic centimetres.
No. of cyl.	Number of cylinders.
Year	Years during which vehicle was manufactured.
Spec.	SU specification number by which complete unit may be ordered, and which should be quoted when ordering spare parts. This number identifies the unit in all respects ie units of the same specification but fitted to different vehicles will be identical with each other and are therefore interchangeable.
Position	Installation position. See Chapter on Installation.
Type	Details type of SU carburettor. In some instances, notably early post-war applications, the carburettor type is not dealt with in this manual, but full details are given in the Applications list for reference.
Needle	Details the needle or needles fitted as standard for any particular application. See Appendix 2, Notes on needles.
Piston spring	Details colour of end coil/s of piston spring (which indicates spring load and rate).

Model Details	Capacity	No. of Cyl.	Year	Spec.	Position	Type	Rich	Needle Std.	Weak	Piston Spring
ALVIS										
TD 21	2993cc	6	1963/64	AUD 128F	F	HD6	–	KA	–	Red
				AUD 128R	R	HD6Th[2]	–	KA	–	Red
TF 21 Series IV	2993cc	6	1965/66	AUD 226F	F	HD6	–	SC	–	Red
				AUD 226C	C	HD6	–	SC	–	Red
				AUD 226R	R	HD6Th[2]	–	SC	–	Red
ASTON MARTIN										
DB5	3670cc	6	1962/64	AUD 88F	F	HD8	–	UX	–	Red/Green
				AUD 88C	C	HD8	–	UX	–	Red/Green
				AUD 88R	R	HD8	–	UX	–	Red/Green
DB6	3995cc	6	1965/67	AUD 88F	F	HD8	–	UX	–	Red/Green
				AUD 88C	C	HD8	–	UX	–	Red/Green
				AUD 88R	R	HD8	–	UX	–	Red/Green
AUSTIN-HEALEY										
Austin-Healey 100 (BN2)	2639cc	4	1953/56		F	H4	QA	QW	AT	Yellow
					R	H4	QA	QW	AT	Yellow
Austin-Healey Le Mans	2639cc	4	1954/56		F	H6	QA6	OA7	OA8	Red
					R	H6	OA6	OA7	OA8	Red
Austin-Healey 100S	2639cc	4	1955		F	H6	KW	KW1	SA	Red
					R	H6	KW	KW1	SA	Red
Austin-Healey 100/6 (BN4)	2639cc	6	1957		F	H4	4	AJ	MI	Red
					R	H4	4	AJ	MI	Red
Austin-Healey 3000 Mk I (BN6)	2912cc	6	1959		F	HD6	RD	CV	SQ	Yellow
					R	HD6	RD	CV	SQ	Yellow
Austin-Healey 3000 Mk II (BN7)	2912cc	6	1959		F	HD6	RD	CV	SQ	Green
					R	HD6	RD	CV	SQ	Green

Model Details	Capacity	No. of Cyl.	Year	Spec.	Position	Type	Needle			Piston Spring
							Rich	Std.	Weak	
Austin Healey 3000 Mk II (BN7)	2912cc	6	1959		F	HD6Th	RD	CV	SQ	Green
					R	HD6	RD	CV	SQ	Green
Austin-Healey 3000 Mk II (BN7) (RC)	2912cc	6	1960		F	HD6	RD	CV	SQ	Green
					R	HD6	RD	CV	SQ	Green
Austin-Healey 3000 Mk II (BN7)	2912cc	6	1961/62		F	HS4	DK	DJ	DH	Red
					C	HS4	DK	DJ	DH	Red
					R	HS4	DK	DJ	DH	Red
Austin-Healey 3000 Mk II	2912cc	6	1962/63		F	HS6	RD	BC	TZ	Green
					R	HS6	RD	BC	TZ	Green
Austin-Healey 3000 Mk III (BJ8)	2912cc	6	1964	AUD 124	F	HD8	UN	UH	UL	Red/Green
				AUD 124	R	HD8	UN	UH	UL	Red/Green
AUSTIN										
A99	2912cc	6	1959/61		F	H4		M5	HA	Yellow
					R	H4		M5	HA	Yellow
Seven (Mini)	848cc	4	1959		–	HS2	M	EB	GG	Red
Seven and Super	848cc	4	1961/62		–	HS2	M	EB	GG	Red
Mini-Cooper	997cc	4	1961/62		F	HS2	AH2	GZ	EB	Red
					R	HS2	AH2	GZ	EB	Red
A40	948cc	4	1961/62		–	HS2	AH2	M	EB	Red
A35 Van	948cc	4	1962/63		–	HS2	H6	AN	EB	Red
A35 Van	848cc	4	1965/70	AUD 120	–	HS2	M	EB	GG	Red
A40 Mk II	1098cc	4	1962/7	AUD 13	–	HS2	H6	AN	EB	Red
Westminster 110	2912cc	6	1967	AUD 240F	F	H4	3	AR	HA	Yellow
				AUD 240R	R	H4	3	AR	HA	Yellow
Westminster A110 Police	2912cc	6	1967	AUD 259F	F	H4	3	AR	HA	Yellow
				AUD 259R	R	H4	3	AR	HA	Yellow
3 litre	2912cc	6	1967/68	AUD 217F	F	HS6	TU	CI	CIW	Yellow
				AUD 217R	R	HS6	TU	CI	CIW	Yellow
A60	1622cc	4	1961/70	AUD 40	–	HS2	M	GX	GG	Yellow
10 cwt Van	1622cc	4	1971/72	AUD 523	–	HS2	M	GX	GG	Yellow
Mini	848cc	4	1962/68	AUC 976	–	HS2	M	EB	GG	Red
Mini Automatic	848cc	4	1965/67	AUD 170	–	HS4	H6	AN	EB	Red
Mini Automatic	848cc	4	1967/68	AUD 250	–	HS4	H6	AN	EB	Red
Mini Mk II	848cc	4	1968/70	AUD 299	–	HS2	M	EB	GG	Red
Mini Mk II Automatic	848cc	4	1969/71	AUD 360	–	HS4	H6	AN	EB	Red
Mini Mk II	848cc	4	1969/74	AUD 369	–	HS2	–	EB	–	Red
Mini Mk II Automatic	848cc	4	1971/74	AUD 394	–	HS4	–	AN	–	Red
Mini (ECE)	848cc	4	1971/74	AUD 449	–	HS2	–	AAV	–	Red
Mini Van (GPO)	848cc	4	1972/74	AUD 587	–	HS2	–	AAV	–	Red
Austin Mini Van (GUS)	848cc	4	1974/	AUD 713	–	HS4	–	ABS	–	Red
Mini (E.C.E.)	848cc	4	1974/	AUD 611	–	HS4	–	ABS	–	Red
Mini Mk II	998cc	4	1967/68	AUD 86	–	HS2	M	GX	GG	Red
Mini Mk II Automatic	998cc	4	1967/68	AUD 184	–	HS4	MI	AC	HA	Red
Mini Mk II	998cc	4	1968/70	AUD 298	–	HS2	M	GX	GG	Red
Mini Mk II Automatic	998 cc	4	1969	AUD 366	–	HS4	MI	AC	HA	Red
Mini Mk II Automatic	998 cc	4	1970	AUD 367	–	HS4	MI	AC	HA	Red
Mini Mk II	998cc	4	1970/71	AUD 363	–	HS2	M	GX	GG	Red
Mini Mk II Automatic	998cc	4	1970/74	AUD 393	–	HS4	MI	AC	HA	Red
Mini Mk II Automatic	998cc	4	1970/74	AUD 393	–	HS4	MI	AC	HA	Red
Mini Mk II (E.C.E.)	998cc	4	1971/74	AUD 509	–	HS2	–	AAV	–	Red
Mini Mk II Man/Auto (ECE)	998cc	4	1974/	AUD 679	–	HS4	–	ABX	–	Red
Mini Van (GPO)	998cc	4	1974/	AUD 706	–	HS4	–	ABX	–	Red
Mini (Canada)	998cc	4	1972/73	AUD 548	–	HS4	–	AAG	–	Red
Mini (Canada)	998cc	4	1973/	AUD 618	–	HS4	–	ABJ	–	Red
Mini (Canada)	998cc	4	1974/	AUD 654	–	HS4	–	ABJ	–	Red
Mini Mk II Man/Auto (ECE) Export only	1098cc	4	1973	AUD 608	–	HS4	–	ABJ	–	Red
Mini Clubman	998cc	4	1969/71	AUD 363	–	HS2	M	GX	GG	Red
Mini Clubman Automatic	998cc	4	1970/74	AUD 393	–	HS4	MI	AC	HA	Red
Mini Clubman (ECE)	998cc	4	1971/74	AUD 509	–	HS2	–	AAV	–	Red
Mini Clubman Man/Auto (ECE)	998cc	4	1974/	AUD 679	–	HS4	–	ABX	–	Red
Mini Clubman Automatic	998cc	4	1972/	AUD 450	–	HS4	–	AAG	–	Red
Mini Clubman 1275 GT	1275cc	4	1969/71	AUD 317	–	HS4	BQ	DZ	CF	Red
Mini Clubman 1275 GT	1275cc	4	1971/72	AUD 451	–	HS4	–	AAR	–	Red
Mini Clubman 1275GT (ECE)	1275cc	4	1972/	AUD 567	–	HS4	–	ABB	–	Red
Mini Cooper Mk I & Mk II	998cc	4	1964/69	AUD 104L	LH	HS2	M	GY	GG	Blue
				AUD 104R	RH	HS2	M	GY	GG	Blue
Mini Cooper S	970cc	4	1964	AUD 151L	LH	HS2	H6	AN	EB	Red
				AUD 151R	RH	HS2	H6	AN	EB	Red
Mini Cooper S	1071cc	4	1963/64	AUD 99L	LH	HS2	3	H6	EB	Red
				AUD 99R	RH	HS2	3	H6	EB	Red
Mini Cooper S	1275cc	4	1964/70	AUD 146L	LH	HS2	AH2	M	EB	Red
				AUD 146R	RH	HS2	AH2	M	EB	Red
Mini Cooper S	1275cc	4	1970/71	AUD 440L	LH	HS2	AH2	M	EB	Red
				AUD 440R	RH	HS2	AH2	M	EB	Red
1100	1098cc	4	1962/67	AUD 13	–	HS2	H6	AN	EB	Red
1100 Automatic	1098cc	4	1965/67	AUD 185	–	HS4	BQ	DL	ED	Red
1100 Mk II	1098cc	4	1967/71	AUD 13	–	HS2	H6	AN	EB	Red
1100 Mk II Automatic	1098cc	4	1967/68	AUD 251	–	HS4	BQ	DL	ED	Red
1100 Mk II Automatic	1098cc	4	1969/71	AUD 370	–	HS4	BQ	DL	ED	Red
1100 Mk II	1098cc	4	1971/72	AUD 368	–	HS2	H6	AN	EB	Red
1100 Mk III	1098cc	4	1971/74	AUD 368	–	HS2	H6	AN	EB	Red
1100 Mk III Automatic	1098cc	4	1971/74	AUD 371	–	HS4	BQ	DL	ED	Red
1100 Mk III (ECE)	1098cc	4	1971/74	AUD 508	–	HS4	–	AAY	–	Red
1300	1275cc	4	1967/68	AUD 186	–	HS4	BQ	DZ	CF	Red
1300 Automatic	1275cc	4	1967/68	AUD 271	–	HS4	BQ	DZ	CF	Red

Model Details	Capacity	No. of Cyl.	Year	Spec.	Position	Type	Rich	Needle Std.	Weak	Piston Spring
1300	1275cc	4	1969/70	AUD 374	–	HS4	BQ	DZ	CF	Red
1300 Automatic	1275cc	4	1969/70	AUD 376	–	HS4	BQ	DZ	CF	Red
1300 GT	1275cc	4	1969/71	AUD 344L	LH	HS2	M	GY	GG	Blue
				AUD 344R	RH	HS2	M	GY	GG	Blue
1300	1275cc	4	1971	AUD 472	–	HS4	BQ	DZ	CF	Red
1300GT	1275cc	4	1971	AUD 431L	LH	HS2	M	GY	GG	Blue
				AUD 431R	RH	HS2	M	GY	GG	Blue
1300	1275cc	4	1971/72	AUD 480	–	HS4	BQ	DZ	CF	Red
1300GT	1275cc	4	1971/72	AUD 454LH	LH	HS2	–	AAP	–	Blue
				AUD 454RH	RH	HS2	–	AAP	–	Blue
1300 Mk I & III (ECE)	1275cc	4	1971/72	AUD 453	–	HS4	–	AAR	–	Red
1300GT (ECE)	1275cc	4	1971/72	AUD 496LH	LH	HS2	–	AAP	–	Blue
				AUD 496RH	RH	HS2	–	AAP	–	Blue
1300 Mk I & III (ECE)	1275cc	4	1972/73	AUD 559	–	HS4	–	ABB	–	Red
1300 Mk III (ECE)	1275cc	4	1972/73	AUD 585	–	HS4	–	AAR	–	Red
1300 Mk III (ECE)	1275cc	4	1973/4	AUD 594	–	HS4	–	ABB	–	Red
1300 Mk III (ECE)	1275cc	4	1973/	AUD 595	–	HS4	–	AAR	–	Red
1300 Mk III Automatic (ECE)	1275cc	4	1971/74	AUD 486	–	HS4	–	AAR	–	Red
1300 Mk III Automatic (ECE)	1275cc	4	1972/74	AUD 567	–	HS4	–	ABB	–	Red
Austin Healey Sprite Mk I	948cc	4	1959		F	H1	EB	GG	MOW	
					R	H1	EB	GG	MOW	
Austin Healey Sprite Mk II	948cc	4	1961/62		F	HS2	V2	V3	GX	Blue
					R	HS2	V2	V3	GX	Blue
Austin Healey Sprite Mk II	1098cc	4	1962/63		F	HS2	M	GY	GG	Blue
					R	HS2	M	GY	GG	Blue
Austin Healey Sprite Mk III	1098cc	4	1964/66	AUD 136F	F	HS2	H6	AN	GG	Blue
				AUD 136R	R	HS2	H6	AN	GG	Blue
Austin Healey Sprite Mk IV	1275cc	4	1967/68	AUD 136F	F	HS2	H6	AN	GG	Blue
				AUD 136R	R	HS2	H6	AN	GG	Blue
Austin Healey Sprite Mk IV	1275cc	4	1968/71	AUD 327F	F	HS2	H6	AN	GG	Blue
				AUD 327R	R	HS2	H6	AN	GG	Blue
Austin Sprite Mk IV	1275cc	4	1971	AUD 327F	F	HS2	H6	AN	GG	Blue
				AUD 327R	R	HS2	H6	AN	GG	Blue
Austin Healey Sprite Mk IV (USA)	1275cc	4	1968	AUD 266F	F	HS2	–	AN	–	Blue
				AUD 266R	R	HS2	–	AN	–	Blue
Austin Healey Sprite Mk IV (USA)	1275cc	4	1968/69	AUD 328F	F	HS2	–	AAC	–	Blue
				AUD 328R	R	HS2	–	AAC	–	Blue
Austin Healey Sprite Mk IV (USA)	1275cc	4	1969/71	AUD 404F	F	HS2	–	AAC	–	Blue
				AUD 404R	R	HS2	–	AAC	–	Blue
Austin Sprite (USA)	1275cc	4	1972/74	AUD 549F	F	HS2	–	ABC	–	Blue
				AUD 549R	R	HS2	–	ABC	–	Blue
Austin America Automatic	1275cc	4	1968	AUD 296	–	HS4	–	DZ	–	Red
Austin America	1275cc	4	1968	AUD 281	–	HS4	–	DZ	–	Red
Austin America Automatic	1275cc	4	1969	AUD 380	–	HS4	–	AAG	–	Red
Austin America	1275cc	4	1969	AUD 379	–	HS4	–	AAG	–	Red
Austin America Automatic	1275cc	4	1969/71	AUD 346	–	HS4	–	AAG	–	Red
Austin America	1275cc	4	1969/71	AUD 345	–	HS4	–	AAG	–	Red
Austin Marina (USA)	1798cc	4	1972	AUD 494	–	HIF6	–	BBE	–	Yellow
Austin Marina Automatic (USA)	1798cc	4	1972	AUD 495	–	HIF6	–	BBE	–	Yellow
Austin Marina (USA)	1798cc	4	1972/74	AUD 583	–	HIF6	–	BBQ	–	Yellow
Austin Marina Automatic (USA)	1798cc	4	1972/74	AUD 584	–	HIF6	–	BBQ	–	Yellow
Austin Marina (Canada)	1798cc	4	1973/	AUD 575	–	HIF6	–	BAG	–	Yellow
Austin Marina Automatic (Canada)	1798cc	4	1973/	AUD 576	–	HIF6	–	BAG	–	Yellow
Austin 7 cwt Van	1098cc	4	1972/73	AUD 368	–	HS2	H6	AN	EB	Red
Austin 7 cwt Van (ECE)	1098cc	4	1973/	AUD 627	–	HS4	–	ABN	–	Red
Austin 10 cwt Van (ECE)	1275cc	4	1972/	AUD 541	–	HS4	–	AAZ	–	Red
Austin 10 cwt GPO Van	1275cc	4	1972/73	AUD 589	–	HS4	–	AAZ	–	Red
Allegro 1100 (ECE)	1098cc	4	1973/	AUD 608	–	HS4	–	ABP	–	Red
Allegro 1300 (ECE)	1275cc	4	1973/	AUD 594	–	HS4	–	ABB	–	Red
Allegro 1300 Automatic (ECE)	1275cc	4	1973/	AUD 567	–	HS4	–	ABB	–	Red
Allegro 1500 (ECE)	1485cc	4	1973/	AUD 556	–	HS6	–	BAS	–	Red
Allegro 1500 Automatic (ECE)	1485cc	4	1973/	AUD 628	–	HS6	–	BAS	–	Red
Allegro 1750 (ECE)	1748cc	4	1973/4	AUD 557	–	HS6	–	BBH	–	Red
Allegro 1750 Automatic (ECE)	1748cc	4	1973/4	AUD 619	–	HS6	–	BBH	–	Red
Allegro HL/Sport (ECE)	1748cc	4	1974/	AUD 539L	LH	HS6	–	BBR	–	Red
				AUD 539R	RH	HS6	–	BBR	–	Red
Maxi 1500	1485cc	4	1969/71	AUD 258	–	HS6	–	KP	–	Red
Maxi 1500	1485cc	4	1971	AUD 468	–	HS6	–	BAS	–	Red
Maxi 1500 (ECE)	1485cc	4	1971/72	AUD 498	–	HS6	–	BAS	–	Red
Maxi 1500	1485cc	4	1972/73	AUD 555	–	HS6	–	BAS	–	Red
Maxi 1500 (ECE)	1485cc	4	1972/	AUD 556	–	HS6	–	BAS	–	Red
Maxi 1750	1748cc	4	1970/71	AUD 462	–	HS6	–	BAR	–	Red
Maxi 1750 Automatic	1748cc	4	1972	AUD 463	–	HS6	–	BAR	–	Red
Maxi 1750 (ECE)	1748cc	4	1971/72	AUD 528	–	HS6	–	BBH	–	Red
Maxi 1750 HL (ECE)	1748cc	4	1972/	AUD 539LH	LH	HS6	–	BBR	–	Red
				AUD 539RH	RH	HS6	–	BBR	–	Red
Maxi 1750	1748cc	4	1972/73	AUD 558	–	HS6	–	BAR	–	Red
Maxi 1750 (ECE)	1748cc	4	1972/	AUD 557	–	HS6	–	BBH	–	Red
Maxi 1750 Automatic (ECE)	1748cc	4	1973/	AUD 619	–	HS6	–	BBH	–	Red
1800	1798cc	4	1964/66	AUD 147	–	HS6	SW	TW	CIW	Yellow
1800	1798cc	4	1966/67	AUD 223	–	HS6	SW	TW	CIW	Yellow

Model Details	Capacity	No. of Cyl.	Year	Spec.	Position	Type	Rich	Needle Std.	Weak	Piston Spring
1800 Mk II	1798cc	4	1968/70	AUD 280	–	HS6	SA	ZH	CIW	Yellow
1800 Mk II Automatic	1798cc	4	1968/70	AUD 291	–	HS6	SA	ZH	CIW	Yellow
1800 Automatic (Canada)	1798cc	4	1968/72	AUD 315	–	HS6	–	BAJ	–	Yellow
1800 (Canada)	1798cc	4	1969/72	AUD 314	–	HS6	–	BAJ	–	Yellow
1800 S	1798cc	4	1969/71	AUD 171LH	LH	HS6	CI	TZ	CIW	Red
				AUD 171RH	RH	HS6	CI	TZ	CIW	Red
1800 Mk II	1798cc	4	1971/72	AUD 524	–	HS6	SA	ZH	CIW	Yellow
1800 Mk II Automatic	1798cc	4	1971/74	AUD 525	–	HS6	SA	ZH	CIW	Yellow
1800 Mk II (ECE)	1798cc	4	1971/72	AUD 355	–	HS6	–	BBF	–	Yellow
1800 Mk II Automatic (ECE)	1798cc	4	1973/	AUD 356	–	HS6	–	BBF	–	Yellow
1800 Mk II (ECE)	1798cc	4	1973/	AUD 564	–	HS6	–	BBF	–	Yellow
1800 Mk II	1798cc	4	1972/73	AUD 565	–	HS6	–	ZH	–	Yellow
1800 Mk II Automatic	1798cc	4	1972/73	AUD 568	–	HS6	–	ZH	–	Yellow
2200	2227cc	6	1972/74	AUD 409F	LH	HS6	–	BBD	–	Red
				AUD 409R	RH	HS6	–	BBD	–	Red
2200 (ECE)	2227cc	6	1972/74	AUD 546F	LH	HIF6	–	BBN	–	Red
				AUD 546R	RH	HIF6	–	BBN	–	Red
2200 Automatic (ECE)	2227cc	6	1972/74	AUD 581F	LH	HIF6	–	BBN	–	Red
				AUD 581R	RH	HIF6	–	BBN	–	Red
BENTLEY										
S3 V8	6230cc	8	1963/64	AUD 54A	RH	HD8	–	US	–	Red/Blue
				AUD 54'B'	LH	HD8	–	US	–	Red/Blue
T Series (SY)	6230cc	8	1965/68	AUD 177 'A'	RH	HD8		UZ	–	Red/Blue
				AUD 177 'B'	LH	HD8	–	UZ	–	Red/Blue
T Series (SY) (USA)	6230cc	8	1968	AUD 269 'A'	RH	HD8	–	UVU	–	Red/Blue
				AUD 269 'B'	LH	HD8	–	UVU	–	Red/Blue
T Series (SY) (USA)	6750cc	8	1969	AUD 389 'A'	RH	HD8	–	BAE	–	Red/Blue
				AUD 389 'B'	LH	HD8	–	BAE	–	Red/Blue
T Series (SY) (USA)	6750cc	8	1969/71	AUD 387 'A'	RH	HD8	–	BAE	–	Red/Blue
				AUD 387 'B'	LH	HD8	–	BAE	–	Red/Blue
CHRYSLER										
Hillman Hunter	1496cc	4	1972/73	AUD 554	–	HS4	–	AAK	–	Green
Hillman Hunter	1725cc	4	1972/73	AUD 554	–	HS4	–	AAK	–	Green
Hillman Avenger 1300	1295cc	4	1973/74	AUD 572	–	HS4C	–	ABR	–	Yellow
Avenger 1300	1295cc	4	1974/	AUD 690	–	HS4C	–	ACA	–	Red
Hillman Avenger 1600	1600cc	4	1973/74	AUD 572	–	HS4C	–	ABR	–	Yellow
Dodge 1800	1800cc	4	1973/	AUD 672	–	HS4C	–	AAU	–	Green
COVENTRY CLIMAX										
FWA Stage I	1098cc	4			F	H4		BE		Blue
					R	H4		BE		Blue
FWA Stage II	1098cc	4			F	H4		BF		Blue
					R	H4		BF		Blue
Lotus Elite	1216cc	4			F	H4		BQ		Blue
					R	H4		BQ		Blue
Lotus Elite	1216cc	4			–	H4		BF		Yellow
FPF	1498cc	4			F	DU6		ZB		
					R	DU6		ZB		
DAIMLER										
SP 250 Sports	2548cc	8	1959/64							
V8 Saloon Automatic	2548cc	8	1964/68	AUD 180L	LH	HD6	TL	TZ	–	Red
				AUD 180R	RH	HD6	TL	TZ	–	Red
V8 Saloon Manual	2548cc	8	1967/68	AUD 180L	LH	HD6	TL	TZ	–	Red
				AUD 180R	RH	HD6	TL	TZ	–	Red
V8 Majestic Major	4561cc	8	1964	AUD 139L	LH	HD8	–	UL	–	Red/Green
				AUD 139R	RH	HD8	–	UL	–	Red/Green
V8 Majestic	4561cc	8	1964	AUD 139L	LH	HD8	–	UL	–	Red/Green
				AUD 139R	RH	HD8	–	UL	–	Red/Green
V8 Majestic Major	4561cc	8	1964/68	AUD 181L	LH	HD8	–	UL	–	Red/Green
				AUD 181R	RH	HD8	–	UL	–	Red/Green
Limousine	4235cc	6	1970/72	AUD 357F	F	HD8Th[2]	–	UM	–	Red/Green
				AUD 357R	R	HD8	–	UM	–	Red/Green
Limousine	4235cc	6	1973/74	AUD 647F	F	HS8 AED[3]	–	BAW	–	Red/Green
				AUD 647R	R	HS8 AED	–	BAW	–	Red/Green
Limousine	4235cc	6	1974/75	AUD 667F	F	HS8 AED[3]	–	BCC	–	Red/Green
				AUD 667R	R	HS8 AED	–	BCC	–	Red/Green
Sovereign	2792cc	6	1968/71	AUD 321F	F	HD8Th[2]	–	UVX	–	Red/Green
				AUD 321R	R	HD8	–	UVX	–	Red/Green
Sovereign (LHD)	2792cc	6	1971/72	AUD 537F	F	HS8 AED[4]	–	BBL	–	Red/Green
				AUD 537R	R	HS8 AED	–	BBL	–	Red/Green
Sovereign	2792cc	6	1971/73	AUD 415F	F	HS8 AED[4]	–	BAU	–	Blue/Black
				AUD 41 5R	R	HS8 AED	–	BAU	–	Blue/Black
Sovereign	4235cc	6	1967/68	AUD 245F	F	HD8 Th[2]	–	UM	–	Red/Green
				AUD 245R	R	HD8	–	UM	–	Red/Green
Sovereign	4235cc	6	1968/71	AUD 357F	F	HD8 Th[2]	–	UM	–	Red/Green
				AUD 357R	R	HD8	–	UM	–	Red/Green
Sovereign (LHD)	4235cc	6	1971/72	AUD 538F	F	HS8 AED[3]	–	BBK	–	Red/Green
				AUD 538R	R	HS8 AED	–	BBK	–	Red/Green
Sovereign	4235cc	6	1971/73	AUD 397F	F	HS8 AED[3]	–	BAW	–	Red/Green
				AUD 397R	R	HS8 AED	–	BAW	–	Red/Green
Sovereign	4235cc	6	1973	AUD 647F	F	HS8 AED[3]	–	BAW	–	Red/Green
				AUD 647R	R	HS8 AED	–	BAW	–	Red/Green
Sovereign	4235cc	6	1973/	AUD 653F	F	HS8 AED[3]	–	BCC	–	Red/Green
				AUD 653R	R	HS8 AED	–	BCC	–	Red/Green

Model Details	Capacity	No. of Cyl.	Year	Spec.	Position	Type	Rich	Needle Std.	Weak	Piston Spring
INNOCENTI										
Mini	848cc	4	1965/66	AUD 210	–	HS2	M	EB	GG	Red
Mini Automatic	848cc	4	1967/68	AUD 262	–	HS4	H6	AN	EB	Red
Mini	998cc	4	1968/69	AUD 324L	LH	HS2	M	GY	GG	Blue
				AUD 324R	RH	HS2	M	GY	GG	Blue
Mini 90	998cc	4	1974/	AUD 693	–	HS4	–	ABB	–	Yellow
Mini Clubman	998cc	4	1970/71	AUD 365L	LH	HS2	M	GY	GG	Blue
				AUD 365R	RH	HS2	M	GY	GG	Blue
Minimatic	998cc	4	1970/71	AUD 460	–	HS4	DL	CZ	GY	Red
Mini 1001 Automatic	998cc	4	1971/74	AUD 513	–	HS4	–	AAR	–	Red
Mini 1300 (ECE)	1275cc	4	1972/	AUD 534L	LH	HS2	–	AAP	–	Blue
				AUD 534R	RH	HS2	–	AAP	–	Blue
Mini 120	1275cc	4	1974/	AUD 692	–	HS6	–	BAU	–	Yellow
1100 IM3	1098cc	4	1963/64	AUD 132L	LH	HS2	D6	D3	GV	Blue
				AUD 132R	RH	HS2	D6	D3	GV	Blue
1100 IM3	1098cc	4	1964/	AUD 160L	LH	HS2	D6	D3	GV	Blue
				AUD 160R	RH	HS2	D6	D3	GV	Blue
1100 IM3	1098cc	4	1964	AUD 168	–	HS2	H6	AN	EB	Red
1100 IM3 Automatic	1098cc	4	1967/68	AUD 263	–	HS4	BQ	DL	ED	Red
1100 IM3	1098cc	4	1970/71	AUD 490L	LH	HS2	M	EB	GG	Blue
				AUD 490R	RH	HS2	M	EB	GG	Blue
1100 IM3	1098cc	4	1971/72	AUD 532L	LH	HS2	M	EB	GG	Blue
				AUD 532R	RH	HS2	M	EB	GG	Blue
Regent 1300	1275cc	4	1974/	AUD 534L	LH	HS2	–	AAP	–	Blue
				AUD 534R	RH	HS2	–	AAP	–	Blue
Regent 1500	1498cc	4	1974/	AUD 633L	LH	HS4	–	ABB	–	Red
				AUD 633R	RH	HS4	–	ABB	–	Red
JAGUAR										
XK 120	3442cc	6	1949/50		F	H6Th	–	RB	–	Red
					R	H6	–	RB	–	Red
XK 120	3442cc	6	1951/54		F	H6Th	53	RF	RG	Red
					R	H6	53	RF	RG	Red
XK 120 (remote air	3442cc	6	1951/54		F	H6Th	WO4	WO2	WO3	Red
cleaner)					R	H6	WO4	WO2	WO3	Red
XK 120 7 : 1 and 8 : 1 C.R.	3442cc	6	1952		F	H8	75	VR	VE	Black/Red
C Type					R	H8	75	VR	VE	Black/Red
XK 120 8 : 1 C.R.	3442cc	6	1952		F	H6Th	–	RG	–	Red
C Type					R	H6	–	RG	–	Red
XK 120 8 : 1 cr	3442cc	6	1952		F	H6Th	–	DG	–	Red
(remote air cleaner)					R	H6	–	DG	–	Red
XK 120 9 : 1 cr	3442cc	6	1952		F	H6Th	–	RC	–	Red
C Type					R	H6	–	RC	–	Red
XK 140 7 : 1 and 8 : 1 cr	3442cc	6	1954		F	H6Th	SA	SJ	LBA	Red
					R	H6	SA	SJ	LBA	
XK 140C 7 : 1 and 8 : 1 cr	3442cc	6	1954		F	H6Th	–	SR	–	Red
(C Type head)					R	H6	–	SR	–	Red
XK 140C 7 : 1 and 8 : 1 cr	3442cc	6	1954		F	H6Th	–	WO2	–	Red
(C Type head) (disc air cleaners)					R	H6				
XK 140C 8 : 1 and 9 : 1 cr	3442cc	6	1954		F	H8	75	VR	VE	Black/Red
(C Type head)					R	H8	75	VR	VE	Black/Red
XK 140C 7 : 1 and 8 : 1 cr	3442cc	6	1955		F	H6Th	–	WO2	–	Red
D/H coupe and standard					R	H6	–	WO2	–	Red
XK 140C 7 : 1 and 8 : 1cr	3442cc	6	1955		F	H6Th	–	WO2	–	Red
RHD F/H Coupe					R	H6	–	WO2	–	Red
XK 140C 7 : 1 and 8 : 1 cr	3442cc	6	1955		F	H6Th	–	WO2	–	Red
LHD F/H Coupe					R	H6	–	WO2	–	Red
XK 140 7 : 1 and 8 : 1 cr	3442cc	6	1955		F	H6Th	SA	SJ	LBA	Red
LHD F/H Coupe					R	H6	SA	SJ	LBA	Red
XK 140 7 : 1 and 8 : 1 cr	3442cc	6	1955		F	H6Th	SA	SJ	LBA	Red
RHD F/H Coupe					R	H6	SA	SJ	LBA	Red
XK 140 7 : 1 and 8 : 1 cr	3442cc	6	1955		F	H6Th	SA	SJ	LBA	Red
Borg-Warner trans.					R	H6	SA	SJ	LBA	Red
XK 140 7 : 1 and 8 : 1 cr	3442cc	6	1956		F	H6Th	SA	SJ	LBA	Red
B-W trans. RHD F/H Coupe					R	H6	SA	SJ	LBA	Red
XK 140 7 : 1 and 8 : 1 cr	3442cc	6	1956		F	H6Th	SA	SJ	LBA	Red
B-W trans. LHD D/H Coupe					R	H6	SA	SJ	LBA	Red
XK 140 7 : 1 and 8 : 1 cr	3442cc	6	1956		F	H6Th	SA	SJ	LBA	Red
B-W trans. RHD D/H Coupe					R	H6	SA	SJ	LBA	Red
XK 150S	3781cc	6	1959/62		F	HD8Th	–	UE	–	Blue/Black
					C	HD8	–	UE	–	Blue/Black
					R	HD8	–	UE	–	Blue/Black
XK 150 3.4 litre	3442cc	6	1959/62		F	HD6Th	WO3	TL	SJ	Red
					R	HD6	WO3	TL	SJ	Red
XK 150 3.8 litre	3781cc	6	1960/62		F	HD6Th	–	TU	–	Red
					R	HD6	–	TU	–	Red
E Type 3.8 litre	3781cc	6	1961/64		F	HD8	–	UM	–	Blue/Black
					C	HD8	–	UM	–	Blue/Black
E Type 4.2 litre	4235cc	6	1965		R	HD8	–	UM	–	Blue/Black
Mk X	3781cc	6	1963/64	AUD 111F	F	HD8Th^2	–	UM	–	Blue/Black
				AUD 111C	C	HD8	–	UM	–	Blue/Black
				AUD 111R	R	HD8	–	UM	–	Blue/Black
Mk X 8 : 1 + 9 : 1 cr	3781cc	6	1964	AUD 144F	F	HD8Th^2	–	UM	–	Blue/Black
				AUD 144C	C	HD8	–	UM	–	Blue/Black
				AUD 144R	R	HD8	–	UM	–	Blue/Black
Mk X Automatic and Overdrive	3781cc	6	1964	AUD 156F	F	HD8Th^2	–	UM	–	Blue/Black
				AUD 156C	C	HD8	–	UM	–	Blue/Black
				AUD 156R	R	HD8	–	UM	–	Blue/Black
Mk X	3781cc	6	1964	AUD 157F	F	HD8Th^2	–	UM	–	Blue/Black
				AUD 157C	C	HD8	–	UM	–	Blue/Black

Model Details	Capacity	No. of Cyl.	Year	Spec.	Position	Type	Rich	Needle Std.	Weak	Piston Spring
				AUD 157R	R	HD8	–	UM	–	Blue/Black
Mk X 8 : 1 & 9 : 1 cr	4235cc	6	1964	AUD 144F	F	HD8Th[2]	–	UM	–	Blue/Black
				AUD 144C	C	HD8	–	UM	–	Blue/Black
				AUD 144R	R	HD8	–	UM	–	Blue/Black
Mk X Automatic & Overdrive	4135cc	6	1964	AUD 156F	F	HD8Th[2]	–	UM	–	Blue/Black
				AUD 156C	C	HD8	–	UM	–	Blue/Black
				AUD 156R	R	HD8	–	UM	–	Blue/Black
Mk X	4235cc	6	1964	AUD 157F	F	HD8Th[2]	–	UM	–	Blue/Black
				AUD 157C	C	HD8	–	UM	–	Blue/Black
				AUD 157R	R	HD8	–	UM	–	Blue/Black
240	2483cc	6	1967/68	AUD 256F	F	HS6	–	TL	–	Red
				AUD 256R	R	HS6	–	TL	–	Red
240 Automatic	2483cc	6	1967/68	AUD 297F	F	HS6	–	TL	–	Red
				AUD 297R	R	HS6	–	TL	–	Red
240	2483cc	6	1968/69	AUD 309F	F	HS6	–	TL	–	Red
				AUD 309R	R	HS6	–	TL	–	Red
240 Automatic	2483cc	6	1968/69	AUD 310F	F	HS6	–	TL	–	Red
				AUD 310R	R	HS6	–	TL	–	Red
3.4 Mk III	3442cc	6	1963/64	AUD 109F	F	HD6Th[2]	–	TL	–	Red
				AUD 109R	R	HD6	–	TL	–	Red
3.4 S Type 8 : 1 + 9 : 1 cr Auto. Man. AC Paper cleaner	3442cc	6	1967/68	AUD 243F	F	HD6Th[2]	–	TL	–	Red
				AUD 243R	R	HD6	–	TL	–	Red
3.8 Mk II	3781cc	6	1963/64	AUD 109F	F	HD6Th[2]	–	TL	–	Red
				AUD 109R	R	HD6	–	TL	–	Red
3.8 7 : 1 cr (Cooper cleaner)	3781cc	6	1964	AUD 155F	F	HD6Th[2]	–	TM	–	Red
				AUD 155R	R	HD6	–	TM	–	Red
3.8 S Type Mk III 8 : 1 + 9 : 1 cr (Paper cleaner)	3781cc	6	1964	AUD 153F	F	HD6Th[2]	–	TL	–	Red
				AUD 153R	R	HD6	–	TL	–	Red
3.8 S Type Mk III 8 : 1 + 9 : 1 cr (Oil bath cleaner)	3781cc	6	1964	AUD 154F	F	HD6Th[2]	–	CI	–	Red
				AUD 154R	R	HD6	–	CI	–	Red
3.8 S Type 8 : 1 and 9 : 1 cr Auto. Man (AC Paper cleaner)	3781cc	6	1967/68	AUD 243F	F	HD6Th[2]	–	TL	–	Red
				AUD 243R	R	HD6	–	TL	–	Red
340 7 : 1 cr Man. Aut. (AC Paper cleaner)	3442cc	6	1967/68	AUD 241F	F	HD6Th[2]	–	TM	–	Red
				AUD 241R	R	HD6	–	TM	–	Red
340 8 : 1 + 9 : 1 cr Man. Aut. (AC Paper cleaner)	3442cc	6	1967/68	AUD 242F	F	HD6Th[2]	–	CI	–	Red
				AUD 242R	R	HD6	–	CI	–	Red
420 Manual 8 : 1 + 9 : 1 cr (AC Paper cleaner)	4235cc	6	1967/68	AUD 239F	F	HD8Th[2]	–	UM	–	Blue/Black
				AUD 239R	R	HD8	–	UM	–	Blue/Black
420 Automatic 8 : 1 + 9 : 1 cr (AC Paper cleaner)	4235cc	6	1967/68	AUD 245F	F	HD8Th[2]	–	UM	–	Red/Green
				AUD 245R	R	HD8	–	UM	–	Red/Green
420G Manual 8 : 1 + 9 : 1 cr (AC Paper cleaner)	4235cc	6	1967/68	AUD 157F	F	HD8Th[2]	–	NA	–	Blue/Black
				AUD 157C	C	HD8	–	NA	–	Blue/Black
				AUD 157R	R	HD8	–	NA	–	Blue/Black
420G Auto. 8 : 1 and 9 : 1 cr (AC Paper cleaner)	4235cc	6	1967/68	AUD 156F	F	HD8Th[2]	–	UM	–	Blue/Black
				AUD 156C	C	HD8	–	UM	–	Blue/Black
				AUD 156R	R	HD8	–	UM	–	Blue/Black
E Type	3781cc	6	1963/64	AUD 112F	F	HD8	–	UM	–	Blue/Black
				AUD 112C	C	HD8	–	UM	–	Blue/Black
				AUD 112R	R	HD8	–	UM	–	Blue/Black
E Type 8 : 1 and 9 : 1 cr	4235cc	6	1967/68	AUD 227F	F	HD8	–	UM	–	Blue/Black
				AUD 227C	C	HD8	–	UM	–	Blue/Black
				AUD 227R	R	HD8	–	UM	–	Blue/Black
E Type 5.3 litre	5343cc	12	1971/73	–	–	–	–	–	–	–
XJ12 5.3 litre	5343cc	12	1972/73	–	–	–	–	–	–	–
2.8 XJ6	2792cc	6	1968/71	AUD 321F	F	HD8Th[2]	–	UVV	–	Blue/Black
				AUD 321R	R	HD8	–	UVV	–	Blue/Black
2.8 XJ6	2792cc	6	1971/72	AUD 415F	F	HS8AED[4]	–	BAU	–	Blue/Black
				AUD 415R	R	HS8AED	–	BAU	–	Blue/Black
2.8 XJ6 (LHD)	2792cc	6	1972/73	AUD 537F	F	HS8AED[4]	–	BBL	–	Red/Green
				AUD 537R	R	HS8AED	–	BBL	–	Red/Green
4.2 XJ6	4235cc	6	1968/71	AUD 357F	F	HD8Th[2]	–	UM	–	Red/Green
				AUD 357R	R	HD8	–	UM	–	Red/Green
4.2 XJ6 (LHD)	4235cc	6	1972/73	AUD 538F	F	HS8AED[3]	–	BBK	–	Red/Green
				AUD 538R	R	HS8AED	–	BBK	–	Red/Green
4.2 XJ6	4235cc	6	1971/73	AUD 397F	F	HS8AED[3]	–	BAW	–	Red/Green
				AUD 397R	R	HS8AED	–	BAW	–	Red/Green
4.2 XJ6	4235cc	6	1973	AUD 647F	F	HS8AED[3]	–	BAW	–	Red/Green
				AUD 647R	R	HS8AED	–	BAW	–	Red/Green
4.2 XJ6	4235cc	6	1973/	AUD 653F	F	HS8AED[3]	–	BCC	–	Red/Green
				AUD 653R	R	HS8AED	–	BCC	–	Red/Green
LEYLAND INTERNATIONAL										
185, 215, 220, Van	1622cc	4	1974/	AUD 620	–	HS6	–	BCT	–	Yellow
215, 220, 240, 250 LC Van	1798cc	4	1974/	AUD 621	–	HS6	–	BCU	–	Yellow
215, 220, 240, 250 LC Auto. Van	1798cc	4	1974/	AUD 658	–	HS6	–	BCU	–	Yellow
Mini (SA)	1097cc	4	1971	AUD 481	–	HS4	DL	CZ	GY	Red
Apache 1300 (SA)	1275cc	4	1970/71	AUD 468	–	HS4	BQ	DZ	CF	Red
Mini GTS (SA)	1275cc	4	1971/	AUD 431L	LH	HS2	–	GY	–	Blue
				AUD 431R	RH	HS2	–	GY	–	Blue

Model Details	Capacity	No. of Cyl.	Year	Spec.	Position	Type	Rich	Needle Std.	Weak	Piston Spring
Apache 1300 (SA)	1275cc	4	1973/74	AUD 595	–	HS4	–	AAR	–	Red
Apache 1300 Automatic (SA)	1275cc	4	1970/74	AUD 317	–	HS4	BQ	DZ	CF	Red
Apache 1300 TC (SA)	1275cc	4	1971/	AUD 431L	LH	HS2	–	GY	–	Blue
				AUD 431R	RH	HS2	–	GY	–	Blue
Marina 1.7 Man/Auto (SA)	1748cc	4	1972	AUD 503	–	HS6	–	BAN	–	Yellow
Marina 2.6 Man/Auto (SA)	2620cc	6	1973/74	AUD 588	–	HS6	–	BCA	–	Green
1500 TC (Aus)	1485cc	4	1968	AUD 385L	LH	HS6	–	KS	–	Blue
				AUD 385R	RH	HS6	–	KS	–	Blue
1500 (Aus)	1485cc	4	1969	AUD 288	–	HS6	–	TD	–	Red
Mini Saloon/Van/Moke (Aus)	1098cc	4	1974	AUD 668	–	HS2	–	ABV	–	Red
Marina 1500 (Aus)	1485cc	4	1972	AUD 487	–	HS4	–	AAF	–	Yellow
Marina 1750 (Aus)	1748cc	4	1972/73	AUD 503	–	HS6	–	BAN	–	Yellow
Marina 1750TC (Aus)	1748cc	4	1972	AUD 504F	F	HS6	–	BAD	–	Red
				AUD 504R	R	HS6	–	BAD	–	Red
Marina 2620 (Aus)	2620cc	6	1973/74	AUD 588	–	HS6	–	BCA	–	Green
1800 Mk II (Aus)	1798cc	4	1968	AUD 381	–	HS6	–	SL	–	Yellow
1800 Mk II Auto (Aus)	1798cc	4	1968	AUD 382	–	HS6	–	SL	–	Yellow
1800 Mk II TC (Aus)	1798cc	4	1968	AUD 385F	LH	HS6	–	KS	–	Blue
				AUD 385R	RH	HS6	–	KS	–	Blue
2200 (Aus)	2227cc	6	1971/72	AUD 419	–	HS6	–	KV	–	Red
P76 (Aus)	2620cc	6	1973/74	AUD 588	–	HS6	–	BCA	–	Green
Mini 850 (Spain)	848cc	4	197[illegible]/74	AUD 449	–	HS2	–	AAV	–	Red
Mini 1000 (Spain)	998cc	4	1971/74	AUD 509	–	HS2	–	AAV	–	Red
Mini GT (Spain)	1275cc	4	1972/74	AUD 559	–	HS4	–	ABB	–	Red
1100 (Spain)	1098cc	4	1971/	AUD 368	–	HS2	–	AN	–	Red
Victoria 1300 (Spain)	1275cc	4	1972/74	AUD 559	–	HS4	–	ABB	–	Red
Victoria 1300TC (Spain)	1275cc	4	–	AUD 496L	LH	HS2	–	AAP	–	Blue
				AUD 496R	RH	HS2	–	AAP	–	Blue
Victoria 1300 (Spain)	1275cc	4	1973/	AUD 593	–	HS6	–	BCH	–	Red
MG										
TF (and 1.5 litre)	1250cc	4	1954/66	–	F	H4	HI	GJ	GL	Blue
					R	H4	HI	GJ	GL	Blue
ZA Magnette	1489cc	4	1954/66	–	F	H2	M	GM	GO	Red
					R	H2	M	GM	GO	Red
MGA	1489cc	4	1955/59	–	F	H4	CC	GS	4	Red
					R	H4	CC	GS	4	Red
ZA/ZB Magnette	1498cc	4	1956/58	–	F	H4	–	EQ	M5	Red
					R	H4	–	EQ	M5	Red
Twin Cam	1588cc	4	1958	–	F	H6	RH	OA6	OA7	Red
					R	H6	RH	OA6	OA7	Red
Magnette III	1489cc	4	1959/61	–	F	HD4	FT	FU	M9	Red
					R	HD4	FT	FU	M9	Red
MGA (Mk I and II)	1588cc	4	1959/62	–	F	H4	RO	6	AO	Red
					R	H4	RO	6	AO	Red
MG Magnette Mk IV	1622cc	4	1961/68	AUD 41F	F	HD4	FU	HB	FK	Red
				AUD 41R	R	HD4	FU	HB	FK	Red
MG 1100	1098cc	4	1962/68	AUD 69L	L	HS2	D6	D3	GY	Blue
				AUD 69R	R	HS2	D6	D3	GY	Blue
Sedan USA	1275cc	4	1967/68	AUD 281	–	HS4	–	DZ	–	Red
Sedan Auto. (USA)	1275cc	4	1963	AUD 296	–	HS4	–	DZ	–	Red
MGC	2912cc	6	1967/68	AUD 150F	F	HS6	SQ	ST	CIW	Yellow
				AUD 150R	R	HS6	SQ	ST	CIW	Yellow
MGC	2912cc	6	1969	AUD 341F	F	HS6	–	ST	–	Yellow
				AUD 341R	R	HS6	–	ST	–	Yellow
MGC (USA)	2912cc	6	1968	AUD 287F	F	HS6	–	KM	–	Yellow
				AUD 287R	R	HS6	–	KM	–	Yellow
MGC (USA)	2912cc	6	1969	AUD 342F	F	HS6	–	BAD	–	Yellow
				AUD 342R	R	HS6	–	BAD	–	Yellow
Midget Mk I	948cc	4	1961/62	–	F	HS2	V2	V3	GX	Blue
					R	HS2	V2	V3	GX	Blue
Midget Mk II	1098cc	4	1962/63	–	F	HS2	M	GY	GG	Blue
					R	HS2	M	GY	GG	Blue
Midget Mk II	1098cc	4	1964	AUD 136F	F	HS2	H6	AN	GG	Blue
				AUD 136R	R	HS2	H6	AN	GG	Blue
Midget Mk III	1275cc	4	1967/68	AUD 136F	F	HS2	H6	AN	GG	Blue
				AUD 136R	R	HS2	H6	AN	GG	Blue
Midget Mk III	1275cc	4	1968/71	AUD 327F	F	HS2	H6	AN	GG	Blue
				AUD 327R	R	HS2	H6	AN	GG	Blue
Midget Mk III	1275cc	4	1971/72	AUD 502F	F	HS2	–	AAT	–	Red
				AUD 502R	R	HS2	–	AAT	–	Red
Midget Mk III (ECE)	1275cc	4	1973/74	AUD 662F	F	HS2	–	AAC	–	Blue
				AUD 662R	R	HS2	–	AAC	–	Blue
Midget Mk III (USA)	1275cc	4	1968	AUD 266F	F	HS2	–	AN	–	Blue
				AUD 266R	R	HS2	–	AN	–	Blue
Midget Mk III (USA)	1275cc	4	1968/69	AUD 328F	F	HS2	–	AAC	–	Blue
				AUD 328R	R	HS2	–	AAC	–	Blue
Midget Mk III (USA)	1275cc	4	1969/70	AUD 404F	F	HS2	–	AAC	–	Blue
				AUD 404R	R	HS2	–	AAC	–	Blue
Midget Mk III (USA)	1275cc	4	1972/74	AUD 549F	F	HS2	–	ABC	–	Blue
				AUD 549R	R	HS2	–	ABC	–	Blue
Midget 1500 (ECE)	1493cc	4	1974/	AUD 665F	F	HS4	–	ABT	–	Red
				AUD 665R	R	HS4	–	ABT	–	Red
MG 1300	1275cc	4	1967	AUD 186	–	HS4	BQ	DZ	CF	Red
MG 1300 Auto.	1275cc	4	1967/68	AUD 271	–	HS4	BQ	DZ	CF	Red
MG 1300	1275cc	4	1969	AUD 318L	LH	HS2	M	EB	GG	Blue
				AUD 318R	RH	HS2	M	EB	GG	Blue
MG 1300	1275cc	4	1969	AUD 374	–	HS4	BQ	DZ	CF	Red
MG 1300 Mk II	1275cc	4	1969/71	AUD 344L	LH	HS2	M	GY	GG	Blue
				AUD 344R	RH	HS2	M	GY	GG	Blue

Model Details	Capacity	No. of Cyl.	Year	Spec.	Position	Type	Rich	Needle Std.	Weak	Piston Spring
MG 1300 Mk II	1275cc	4	1971	AUD 431L	LH	HS2	M	GY	GG	Blue
				AUD 431R	RH	HS2	M	GY	GG	Blue
MG 1300 Mk II	1275cc	4	1971/72	AUD 454LH	LH	HS2	–	AAP	–	Blue
				AUD 454RH	RH	HS2	–	AAP	–	Blue
MG 1300 Mk II (ECE)	1275cc	4	1971/72	AUD 496L	LH	HS2	–	AAP	–	Blue
				AUD 496R	RH	HS2	–	AAP	–	Blue
MGB	1798cc	4	1962/63		F	HS4	6	MB	21	Red
					R	HS4	6	MB	21	Red
MGB (USA)	1798cc	4	1968	AUD 265F	F	HS4	–	FX	–	Red
				AUD 265R	R	HS4	–	FX	–	Red
MGB Mk II (USA)	1798cc	4	1968/69	AUD 326F	F	HS4	–	AAE	–	Red
				AUD 326R	R	HS4	–	AAE	–	Red
MGB Mk II (USA)	1798cc	4	1970/71	AUD 405F	F	HS4	–	AAE	–	Red
				AUD 405R	R	HS4	–	AAE	–	Red
MGB Mk II (USA)	1798cc	4	1971	AUD 465F	F	HS4	–	AAL	–	Red
				AUD 465R	R	HS4	–	AAL	–	Red
MGB Mk II (USA)	1798cc	4	1972	AUD 493F	F	HIF4	–	AAU	–	Red
				AUD 493R	R	HIF4	–	AAU	–	Red
MGB	1798cc	4	1972/74	AUD 550F	F	HIF4	–	ABD	–	Red
				AUD 550R	R	HIF4	–	ABD	–	Red
MGB (USA)	1798cc	4	1974/	AUD 630F	F	HIF4	–	ABD	–	Red
				AUD 630R	R	HIF4	–	ABD	–	Red
MGB Competition	1798cc	4	1963/64	AUD 129F	F	HD8	–	UVD	–	Blue/Black
				AUD 129R	R	HD8	–	UVD	–	Blue/Black
MGB & GT	1798cc	4	1965/66	AUD 135F	F	HS4	6	5	21	Red
				AUD 135R	R	HS4	6	5	21	Red
MGB & GT	1798cc	4	1967/68	AUD 278F	F	HS4	5	FX	GZ	Red
				AUD 278R	R	HS4	5	FX	GZ	Red
MGB	1798cc	4	1969/71	AUD 325F	F	HS4	5	FX	GZ	Red
				AUD 325R	R	HS4	5	FX	GZ	Red
MGB	1798cc	4	1972	AUD 434F	F	HIF4	–	AAU	–	Red
				AUD 434R	R	HIF4	–	AAU	–	Red
MGB (ECE)	1798cc	4	1973/74	AUD 616F	F	HIF4	–	AAU	–	Red
				AUD 616R	R	HIF4	–	AAU	–	Red
MGB (ECE)	1798cc	4	1974/	FZX 1001F	F	HIF4	–	ACD	–	Red
				FZX 1001R	R	HIF4	–	ACD	–	Red
MGB GT V8 (ECE)	3528cc	8	1973/	AUD 613L	LH	HIF6	–	BBU	–	Yellow
				AUD 613R	RH	HIF6	–	BBU	–	Yellow
MORRIS										
Minor (Series II) ohv	800cc	4	1953/56	–	–	H1	EB	GG	MOW	
Minor 1000	948cc	4	1957	–	–	H2	S	BX1	MO	Red
Minor 1000 (paper air cleaner)	948cc	4	1957	–	–	H2	AH2	M	EB	Red
Minor 1000 (Rubber fuel line)	948cc	4	1957	–	–	H2	S	BX1	MO	Red
Minor 1000 (Rubber fuel line and paper air cleaner)	948cc	4	1957	–	–	H2	–	M	–	Red
Minor 1000 (Steel levers)	948cc	4	1958	–	–	H2	S	BX1	MO	Red
Minor 1000 (Steel levers and Paper air cleaner)	948cc	4	1958/59	–	–	H2	AH2	M	EB	Red
Mini Minor	848cc	4	1959/62	–	–	HS2	M	EB	GG	Red
Minor 1000	948cc	4	1960/62	–	–	HS2	AH2	M	EB	Red
Minor and 1100	1098cc	4	1962/63	–	–	HS2	H6	AN	EB	Red
Minor	1098cc	4	1962/70	AUD 13	–	HS2	H6	AN	EB	Red
Oxford	1622cc	4	1961/71	AUD 40	–	HS2	M	GX	GG	Yellow
Mini	848cc	4	1962/68	AUC 976	–	HS2	M	EB	GG	Red
Mini Automatic	848cc	4	1965/66	AUD 170	–	HS4	H6	AN	EB	Red
Mini Automatic	848cc	4	1967	AUD 250	–	HS4	H6	AN	EB	Red
Mini	848cc	4	1968/71	AUD 299	–	HS2	M	EB	GG	Red
Mini Automatic	848cc	4	1969	AUD 360	–	HS4	H6	AN	EB	Red
Mini Mk II	848cc	4	1969/72	AUD 359	–	HS2	–	EB	–	Red
Mini (ECE)	848cc	4	1971/74	AUD 449	–	HS2	–	AAV	–	Red
Mini Van (GPO)	848cc	4	1972/73	AUD 587	–	HS2	–	AAV	–	Red
Mini Mk II	998cc	4	1967/68	AUD 86	–	HS2	M	GX	GG	Red
Mini Mk II Auto	998cc	4	1967/68	AUD 184	–	HS4	MI	AC	HA	Red
Mini Mk II	998cc	4	1968/71	AUD 298	–	HS2	M	GX	GG	Red
Mini Mk II Auto	998cc	4	1969	AUD 366	–	HS4	MI	AC	HA	Red
Mini Mk II	998cc	4	1969	AUD 363	–	HS2	M	GX	GG	Red
Mini Mk II Auto	998cc	4	1970	AUD 367	–	HS4	MI	AC	HA	Red
Mini Mk II Auto	998cc	4	1970/74	AUD 393	–	HS4	MI	AC	HA	Red
Mini ECE	998cc	4	1971/74	AUD 509	–	HS2	–	AAV	–	Red
Mini Mk II Auto (ECE)	998cc	4	1974/	AUD 679	–	HS4	–	ABX	–	Red
Mini Clubman	998cc	4	1969	AUD 363	–	HS2	M	GX	GG	Red
Mini Clubman Auto	998cc	4	1970/74	AUD 393	–	HS4	MI	AC	HA	Red
Mini Clubman (ECE)	998cc	4	1971/74	AUD 509	–	HS2	–	AAV	–	Red
Mini Clubman Auto	998cc	4	1972/	AUD 450	–	HS4	–	AAG	–	Red
Mini Clubman Auto/Man (ECE)	998cc	4	1974/	AUD 679	–	HS4	–	ABX	–	Red
Mini Van (GPO)	998cc	4	1974/	AUD 706	–	HS4	–	ABX	–	Red
Mini Mk II Auto/Man (ECE) Expt. only	1098cc	4	1973/	AUD 608	–	HS4	–	ABP	–	Red
Mini Clubman 1275 GT	1275cc	4	1969	AUD 317	–	HS4	BQ	DZ	CF	Red
Mini Clubman 1275 GT (ECE)	1275cc	4	1971/72	AUD 451	–	HS4	–	AAR	–	Red
Mini Clubman 1275 GT (ECE)	1275cc	4	1972/	AUD 567	–	HS4	–	ABB	–	Red
Mini Cooper	998cc	4	1964/69	AUD 104L	L	HS2	M	GY	GG	Blue
				AUD 104R	R	HS2	M	GY	GG	Blue
Mini Cooper S	970cc	4	1964	AUD 151L	L	HS2	H6	AN	EB	Red
				AUD 151R	R	HS2	H6	AN	EB	Red

Model Details	Capacity	No. of Cyl.	Year	Spec.	Position	Type	Rich	Needle Std.	Weak	Piston Spring
Mini Cooper S	1071cc	4	1963/64	AUD 99L	L	HS2	3	H6	EB	Red
				AUD 99R	R	HS2	3	H6	EB	Red
Mini Cooper S	1275cc	4	1964/70	AUD 146L	L	HS2	AH2	M	EB	Red
				AUD 146R	R	HS2	AH2	M	EB	Red
Mini Cooper S	1275cc	4	1970/71	AUD 440L	L	HS2	AH2	M	EB	Red
				AUD 440R	R	HS2	AH2	M	EB	Red
1100 Auto	1098cc	4	1965/66	AUD 185	–	HS4	BQ	DL	ED	Red
1100 Auto	1098cc	4	1967	AUD 251	–	HS4	BQ	DL	ED	Red
1100 Mk II	1098cc	4	1967/68	AUD 13	–	HS2	H6	AN	EB	Red
1100 Mk II Auto	1098cc	4	1969/71	AUD 370	–	HS4	BQ	DL	ED	Red
1300	1275cc	4	1967/68	AUD 186	–	HS4	BQ	DZ	CF	Red
1300 Auto	1275cc	4	1967/68	AUD 271	–	HS4	BQ	DZ	CF	Red
1300	1275cc	4	1969/70	AUD 374	–	HS4	BQ	DZ	CF	Red
1300 Auto	1275cc	4	1969/70	AUD 376	–	HS4	BQ	DZ	CF	Red
1300 GT	1275cc	4	1969/71	AUD 344L	L	HS2	M	GY	GG	Blue
				AUD 344R	R	HS2	M	GY	GG	Blue
1300	1275cc	4	1971	AUD 472	–	HS4	BQ	DZ	CF	Red
1300 GT	1275cc	4	1971	AUD 431L	L	HS2	M	GY	GG	Blue
				AUD 431R	R	HS2	M	GY	GG	Blue
1300	1275cc	4	1971	AUD 480	–	HS4	BQ	DZ	CF	Red
1300 Traveller (ECE)	1275cc	4	1971/72	AUD 453	–	HS4	–	AAR	–	Red
1300 Traveller	1275cc	4	1972/73	AUD 559	–	HS4	–	ABB	–	Red
1300 Mk III Traveller	1275cc	4	1972/73	AUD 585	–	HS4	–	AAR	–	Red
1300 Mk III Traveller (ECE)	1275cc	4	1973/74	AUD 594	–	HS4	–	ABB	–	Red
1300 Mk III Traveller	1275cc	4	1973	AUD 595	–	HS4	–	AAR	–	Red
1300 Traveller Auto (ECE)	1275cc	4	1971/72	AUD 486	–	HS4	–	AAR	–	Red
1300 Mk III Traveller Auto (ECE)	1275cc	4	1972/74	AUD 567	–	HS4	–	ABB	–	Red
Marina 1.3	1275cc	4	1971/72	AUD 354	–	HS4	–	AAQ	–	Red
Marina 1.3 Auto	1275cc	4	1971/72	AUD 436	–	HS4	–	AAQ	–	Red
Marina 1.3 (ECE)	1275cc	4	1972/	AUD 541	–	HS4	–	AAZ	–	Red
Marina 1.3 Auto (ECE)	1275cc	4	1972/	AUD 542	–	HS4	–	AAZ	–	Red
Marina 1.8	1798cc	4	1971/72	AUD 428	–	HS6	–	BAQ	–	Yellow
Marina 1.8 Auto	1798cc	4	1971/72	AUD 479	–	HS6	–	BAQ	–	Yellow
Marina 1.8 TC	1798cc	4	1971/72	AUD 445F	F	HS4	–	AAS	–	Red
				AUD 445R	R	HS4	–	AAS	–	Red
Marina 1.8 TC Auto	1798cc	4	1971/72	AUD 464F	F	HS4	–	AAS	–	Red
				AUD 464R	R	HS4	–	AAS	–	Red
Marina 1.8 (ECE)	1798cc	4	1972/74	AUD 535	–	HS6	–	BAS	–	Yellow
Marina 1.8 Auto (ECE)	1798cc	4	1972/74	AUD 536	–	HS6	–	BAS	–	Yellow
Marina 1.8 TC (ECE)	1798cc	4	1972/74	AUD 543F	F	HS4	–	ABA	–	Red
				AUD 543R	R	HS4	–	ABA	–	Red
Marina 1.8 (ECE)	1798cc	4	1972/74	AUD 566	–	HS6	–	BAS	–	Yellow
Marina 1.8 TC (ECE) Auto	1798cc	4	1973/74	AUD 573F	F	HS4	–	AAS	–	Red
				AUD 673R	R	HS4	–	AAS	–	Red
Marina 1.8 (ECE)	1798cc	4	1974/	FZX 1011	–	HS6	–	BCW	–	Yellow
Marina 1.8 Auto (ECE)	1798cc	4	1974/	FZX 1012	–	HS6	–	BCW	–	Yellow
Marina 1.8 TC (ECE)	1798cc	4	1974/	FZX 1013F	F	HS4	–	ACE	–	Red
				FZX 1013R	R	HS4	–	ACE	–	Red
Marina 1.8 TC Auto (ECE)	1798cc	4	1974/	FZX 1014F	F	HS4	–	ACE	–	Red
				FZX 1014R	R	HS4	–	ACE	–	Red
7 cwt Van	1098cc	4	1972/73	AUD 268	–	HS2	–	AN	–	Red
7 cwt Van (ECE)	1098cc	4	1973/	AUD 627	–	HS4	–	ABN	–	Red
10 cwt Van (ECE)	1275cc	4	1972/	AUD 541	–	HS4	–	AAZ	–	Red
10 cwt GPO Van	1275cc	4	1972/73	AUD 589	–	HS4	–	AAZ	–	Red
1800	1798cc	4	1964	AUD 147	–	HS6	SW	TW	CIW	Yellow
1800	1798cc	4	1966	AUD 223	–	HS6	SW	TW	CIW	Yellow
1800 Mk II	1798cc	4	1968	AUD 280	–	HS6	SA	ZH	CIW	Yellow
1800 Mk II Auto	1798cc	4	1968	AUD 291	–	HS6	SA	ZH	CIW	Yellow
1800 S	1798cc	4	1969/71	AUD 171L	LH	HS6	CI	TZ	CIW	Red
				AUD 171R	RH	HS6	CI	TZ	CIW	Red
1800 Mk II	1798cc	4	1971/72	AUD 524	–	HS6	SA	ZH	CIW	Yellow
1800 Mk II Auto	1798cc	4	1971/72	AUD 525	–	HS6	SA	ZH	CIW	Yellow
1800 Mk II (ECE)	1798cc	4	1971/72	AUD 355	–	HS6	–	BBF	–	Yellow
1800 Mk II (ECE)	1798cc	4	1973/74	AUD 564	–	HS6	–	BBF	–	Yellow
1800 Mk II	1798cc	4	1972/73	AUD 565	–	HS6	–	ZH	–	Yellow
1800 Mk II Auto	1798cc	4	1972/73	AUD 568	–	HS6	–	ZH	–	Yellow
1800 Mk II Auto (ECE)	1798cc	4	1971/74	AUD 356	–	HS6	–	BBF	–	Yellow
2200	2227cc	6	1972/74	AUD 409F	LH	HS6	–	BBD	–	Red
				AUD 409R	RH	HS6	–	BBD	–	Red
2200 (ECE)	2227cc	6	1972/74	AUD 546F	LH	HIF6	–	BBN	–	Red
				AUD 546R	RH	HIF6	–	BBN	–	Red
2200 Auto (ECE)	2227cc	6	1972/74	AUD 581F	LH	HIF6	–	BBN	–	Red
				AUD 581R	RH	HIF6	–	BBN	–	Red
LAND ROVER										
2.6 109 FWD. Forward Control	2625cc	6	1963/67	AUD 816	–	HD6	–	SS	–	Yellow
2.6 Station Wagon 109WB (LC)	2625cc	6	1967	AUD 2476	–	HD6	–	SS	–	Yellow
2.6 109 WB (LHD)	2625cc	6	1967/68	AUD 2012	–	HD8	–	UG	–	Red/Green
RELIANT										
Sabre-Ford	1703cc	4	1962/63	–	F	HS4	–	CZ	–	Red
					R	HS4	–	CZ	–	Red
Sabre-Ford	1703cc	4	1963/64	AUD 118F	F	HS4	–	DH	–	Red
				AUD 118R	R	HS4	–	DH	–	Red
Scimitar (in line)	2553cc	6	1965/66	AUD 161F	F	HS4	–	GE/R	–	Red
				AUD 161C	C	HS4	–	GE/R	–	Red
				AUD 161R	R	HS4	–	GE/R	–	Red

Model Details	Capacity	No. of Cyl.	Year	Spec.	Position	Type	Rich	Needle Std.	Weak	Piston Spring
RILEY										
4/68	1489cc	4	1959/61		F	HD4	FT	FU	M9	Red
					R	HD4	FT	FU	M9	Red
Elf	848cc	4	1961/62		–	HS2	M	EB	GG	Red
One-Point-Five	1498cc	4	1957/64	AUC 864F	F	H4	AR	AD	HA	Red
				AUC 864R	R	H4	AR	AD	HA	Red
4/72 Saloon	1622cc	4	1961/69	AUD 41F	F	HD4	FU	HB	FK	Red
				AUD 41R	R	HD4	FU	HB	FK	Red
Elf Mk II	998cc	4	1963/64	AUD 86	–	HS2	M	GX	GG	Red
Elf Mk III	998cc	4	1968/69	AUD 298	–	HS2	M	GX	GG	Red
Kestrel	1098cc	4	1965/66	AUD 69L	LH	HS2	D6	D3	GV	Blue
				AUD 69R	RH	HS2	D6	D3	GV	Blue
Kestrel	1275cc	4	1967/68	AUD 186	–	HS4	BQ	DZ	CF	Red
Kestrel Auto	1275cc	4	1967/68	AUD 271	–	HS4	BQ	DZ	CF	Red
Kestrel Mk II	1275cc	4	1968	AUD 318L	LH	HS2	M	EB	GG	Blue
				AUD 318R	RH	HS2	M	EB	GG	Blue
Kestrel Mk II	1275cc	4	1968/69	AUD 344L	LH	HS2	M	GY	GG	Blue
				AUD 344R	RH	HS2	M	GY	GG	Blue
ROLLS-ROYCE										
B61 Power Unit	4887cc	6	1964/69	AUD 55F	F	HS6	CIW	TV	–	Green
				AUD 55R	R	HS6	CIW	TV	–	Green
B61 Power Unit	4887cc	6	1971	AUD 477F	F	HS6	–	CIW	–	Green
				AUD 477R	R	HS6	–	CIW	–	Green
B61 Power Unit	4887cc	6	1974/	AUD 646F	F	HS8	–	BAH	–	Red/Green
				AUD 646R	R	HS8	–	BAH	–	Red/Green
Phantom V	6230cc	8	1969	AUD 384A	RH	HD8	–	US	–	Red/Blue
				AUD 384B	LH	HD8	–	US	–	Red/Blue
Phantom VI	6230cc	8	1971/72	AUD 474A	RH	HD8	–	BAM	–	Red/Blue
				AUD 474B	LH	HD8	–	BAM	–	Red/Blue
Phantom VI	6230cc	8	1971/72	AUD 446A	RH	HD8	–	US	–	Red/Blue
				AUD 446B	LH	HD8	–	US	–	Red/Blue
Phantom VI	6230cc	8	1973/	AUD 656A	RH	HD8	–	BCE	–	Red/Blue
				AUD 656B	LH	HD8	–	BCE	–	Red/Blue
S3 V8	6230cc	8	1963/64	AUD 54A	RH	HD8	–	US	–	Red/Blue
				AUD 54B	LH	HD8	–	US	–	Red/Blue
Silver Shadow	6230cc	8	1965/68	AUD 177A	RH	HD8	–	UZ	–	Red/Blue
				AUD 177B	LH	HD8	–	UZ	–	Red/Blue
Silver Shadow (USA)	6230cc	8	1968	AUD 269A	RH	HD8	–	UVU	–	Red/Blue
				AUD 269B	LH	HD8	–	UVU	–	Red/Blue
Silver Shadow (USA)	6750cc	8	1969	AUD 389A	RH	HD8	–	BAE	–	Red/Blue
				AUD 389B	LH	HD8	–	BAE	–	Red/Blue
Silver Shadow (USA & General)	6750cc	8	1969/71	AUD 387A	RH	HD8	–	BAE	–	Red/Blue
				AUD 387B	LH	HD8	–	BAE	–	Red/Blue
Silver Shadow (USA & General)	6750cc	8	1972	AUD 526A	RH	HD8	–	BAE	–	Red/Blue
				AUD 526B	LH	HD8	–	BAE	–	Red/Blue
Silver Shadow (Home Market & Europe)	6750cc	8	1973	AUD 526A	RH	HD8	–	BAE	–	Red/Blue
				AUD 526B	LH	HD8	–	BAE	–	Red/Blue
Silver Shadow/Corniche (USA)	6750cc	8	1973	AUD 574A	RH	HD8	–	BBS	–	Red/Blue
				AUD 574B	LH	HD8	–	BBS	–	Red/Blue
Silver Shadow (Japan)	6750cc	8	1973/	AUD 671A	RH	HD8	–	BBY	–	Red/Blue
				AUD 671B	LH	HD8	–	BBY	–	Red/Blue
Silver Shadow/Corniche (USA)	6750cc	8	1974	AUD 648A	RH	HD8	–	BCB	–	Red/Blue
				AUD 648B	LH	HD8	–	BCB	–	Red/Blue
Corniche	6750cc	8	1971	AUD 474A	RH	HD8	–	BAM	–	Red/Blue
				AUD 474B	LH	HD8	–	BAM	–	Red/Blue
Corniche (Home Market & Europe)	6750cc	8	1972/	AUD 530A	RH	HD8	–	BAM	–	Red/Blue
				AUD 530B	LH	HD8	–	BAM	–	Red/Blue
Silver Shadow/Corniche (USA)	6750cc	8	1974/	AUD 702A	RH	HD8	–	BCQ	–	Red/Blue
				AUD 702B	LH	HD8	–	BCQ	–	Red/Blue
ROVER										
3 litre Coupe P5	2995cc	6	1963/64	AUC 982	–	HD8[6]	–	UR	–	Red/Green
3 litre P5	2995cc	6	1963/64	AUD 114	–	HD8[6]	–	UR	–	Red/Green
3 litre	2995cc	6	1963/64	AUD 115	–	HD8[6]	–	UR	–	Red/Green
2000	1975cc	4	1963/64	AUC 968	–	HS6	–	RN	–	Green
2000	1975cc	4	1963/64	AUD 141	–	HS6	–	RR	–	Green
2000	1975cc	4	1965/68	AUD 211	–	HS6	–	RN	–	Green
2000	1975cc	4	1969/71	AUD 401	–	HS6	–	KU	–	Green
2000 (ECE)	1975cc	4	1971	AUD 475	–	HS6	–	BAF	–	Green
2000 TC	1975cc	4	1966	AUD 92F	F	HD8	–	UI	–	Blue/Black
				AUD 92R	R	HD8	–	UI	–	Blue/Black
2000 TC	1975cc	4	1967/68	AUD 264F	F	HS8	–	AAA	–	Blue/Black
				AUD 264R	R	HS8	–	AAA	–	Blue/Black
2000 TC	1975cc	4	1969/71	AUD 330F	F	HS8	–	AAA	–	Blue/Black
				AUD 330R	R	HS8	–	AAA	–	Blue/Black
2000 TC	1975cc	4	1971/73	AUD 533F	F	HS8	–	AAM	–	Blue/Black
				AUD 533R	R	HS8	–	AAM	–	Blue/Black
2000 (USA)	1975cc	4	1967/68	AUD 267	–	HS6	–	RR	–	Green
2000 TC (USA)	1975cc	4	1967/68	AUD 254F	F	HS8	–	AAA	–	Blue/Black
				AUD 254R	R	HS8	–	AAA	–	Blue/Black
2000 TC (USA)	1975cc	4	1968	AUD 329F	F	HS8	–	AAB	–	Blue/Black
				AUD 329R	R	HS8	–	AAB	–	Blue/Black
2000 TC (USA ECE)	1975cc	4	1969/74	AUD 411F	F	HS8	–	AAB	–	Blue/Black
				AUD 411R	R	HS8	–	AAB	–	Blue/Black

Model Details	Capacity	No. of Cyl.	Year	Spec.	Position	Type	Rich	Needle Std.	Weak	Piston Spring
2200 SC	2204cc	4	1973/	AUD 631	–	HIF6	–	BBW	–	Green
2200 TC	2204cc	4	1973/	AUD 632F	F	HIF6	–	BBX	–	Yellow
				AUD 632R	R	HIF6	–	BBX	–	Yellow
3.5 litre V8 P5	3528cc	8	1967/68	AUD 233L	LH	HS6	–	KL	–	Yellow
				AUD 233R	RH	HS6	–	KL	–	Yellow
3.5 litre V8 P5	3528cc	8	1968/69	AUD 270L	LH	HS6AED[5]	–	KL	–	Yellow
				AUD 270R	RH	HS6AED	–	KL	–	Yellow
3.5 litre V8 P6	3528cc	8	1968	AUD 313L	LH	HS6	–	KO	–	Yellow
				AUD 313R	RH	HS6	–	KO	–	Yellow
3500 V8 P6	3528cc	8	1968	AUD 350L	LH	HS6AED[5]	–	KO	–	Yellow
				AUD 350R	RH	HS6AED	–	KO	–	Yellow
3500S V8 P6 (USA)	3528cc	8	1969/70	AUD 312L	LH	HS6	–	BAC	–	Yellow
				AUD 312R	RH	HS6	–	BAC	–	Yellow
3500S P6 (USA)	3528cc	8	1969/70	AUD 412L	LH	HS6AED[8]	–	BAC	–	Yellow
				AUD 412R	RH	HS6AED	–	BAC	–	Yellow
3500 V8 P6	3528cc	8	1971/72	AUD 467L	LH	HS6	–	BAK	–	Yellow
				AUD 467R	RH	HS6	–	BAK	–	Yellow
3500 V8 P6 (ECE)	3528cc	8	1972/73	AUD 408L	LH	HIF6	–	BBG	–	Yellow
				AUD 408R	RH	HIF6	–	BBG	–	Yellow
3500 V8 P6	3528cc	8	1972/73	AUD 521L	LH	HIF6	–	BBG	–	Yellow
				AUD 521R	RH	HIF6	–	BBG	–	Yellow
3500 V8 P6 (ECE) & 3500S V8 P6 (ECE)	3528cc	8	1973/76	AUD 623L	LH	HIF6	–	BBV	–	Yellow
				AUD 623R	RH	HIF6	–	BBV	–	Yellow
3500 (Japan)	3528cc	8	1973/76	AUD 669L	LH	HIF6	–	BBV	–	Yellow
				AUD 669R	RH	HIF6	–	BBV	–	Yellow
STANDARD										
8 h.p. and 10 h.p.	803cc	4	1955/56	–	F	H1	–	D3	–	
	948cc	4	1955/56	–	R	H1	–	D3	–	
TRIUMPH										
TR2	1991cc	4	1953/55		F	H4	GER	FV	CR	Red
					R	H4	GER	FV	CR	Red
Herald	948cc	4	1959/61		F	H1	EB	GV	CA	
					R	H1	EB	CV	CA	
TR3, TR3A and TR4	1991cc	4	1956/62		F	H6	RH	SM	SL	Red
					R	H6	RH	SM	SL	Red
TR4A	2138cc	4	1965/66	AUD 209F	F	HS6	SW	TW	CIW	Red
				AUD 209R	R	HS6	SW	TW	CIW	Red
TR4A (USA)	2138cc	4	1968	AUD 284F	F	HS6	–	QW	–	Red
				AUD 284R	R	HS6	–	QW	–	Red
Spitfire Mk I & II	950cc	4	1962/66	AUC 983F	F	HS2	H6	AN	EB	Red
				AUC 983R	R	HS2	H6	AN	EB	Red
Spitfire Mk III	1298cc	4	1967/70	AUD 257F	F	HS2	–	BO	–	Red
				AUD 257R	R	HS2	–	BO	–	Red
Spitfire Mk III (USA)	1296cc	4	1969	AUD 285F	F	HS2	–	DD	–	Red
				AUD 285R	R	HS2	–	DD	–	Red
Spitfire Mk III	1296cc	4	1967/68	AUD 275F	F	HS2	–	DD	–	Red
				AUD 275R	R	HS2	–	DD	–	Red
Spitfire Mk III (USA)	1296cc	4	1967/68	AUD 290F	F	HS2	–	DD	–	Red
				AUD 290R	R	HS2	–	DD	–	Red
Spitfire Mk IV	1296cc	4	1970/71	AUD 441F	F	HS2	–	AAN	–	Red
				AUD 441R	R	HS2	–	AAN	–	Red
Spitfire Mk IV (ECE)	1296cc	4	1972	AUD 517F	F	HS2	–	AAN	–	Red
				AUD 517R	R	HS2	–	AAN	–	Red
Spitfire Mk IV	1296cc	4	1973	AUD 580F	F	HS2	–	AAN	–	Red
				AUD 580R	R	HS2	–	AAN	–	Red
Spitfire Mk V (ECE)	1296cc	4	1973/	AUD 624F	F	HS2	–	AAN	–	Red
				AUD 624R	R	HS2	–	AAN	–	Red
Spitfire 1500 (ECE)	1493cc	4	1974/	AUD 665F	F	HS4	–	ABT	–	Red
				AUD 665R	R	HS4	–	ABT	–	Red
1300 TC	1295cc	4	1967/68	AUD 257F	F	HS2	–	BO	–	Red
				AUD 257R	R	HS2	–	BO	–	Red
Toledo	1296cc	4	1970/71	AUD 392	–	HS4	–	AAK	–	Red
Toledo (ECE)	1296cc	4	1972	AUD 516	–	HS4	–	AAW	–	Red
Toledo	1296cc	4	1972/74	AUD 577	–	HS4	–	ABF	–	Red
Toledo TS	1493cc	4	1974/	AUD 665F	F	HS4	–	ABT	–	Red
				AUD 665R	R	HS4	–	ABT	–	Red
1500	1493cc	4	1970/71	AUD 392	–	HS4	–	AAK	–	Red
1500 (ECE)	1493cc	4	1972/73	AUD 516	–	HS4	–	AAK	–	Red
1500	1493cc	4	1972/74	AUD 578	–	HS4	–	ABG	–	Red
1500 TC (ECE)	1493cc	4	1972/73	AUD 519F	F	HS2	–	AAX	–	Red
				AUD 519R	R	HS2	–	AAX	–	Red
1500 (ECE)	1493cc	4	1973	AUD 579	–	HS4	–	ABG	–	Red
1500 TC	1493cc	4	1973	AUD 582F	F	HS2	–	AAX	–	Red
				AUD 582R	R	HS2	–	AAX	–	Red
1500 TC (ECE)	1493cc	4	1973/74	AUD 625F	F	HS2	–	AAX	–	Red
				AUD 625R	R	HS2	–	AAX	–	Red
1500	1493cc	4	1974/	AUD 665F	F	HS4	–	ABT	–	Red
				AUD 665R	R	HS4	–	ABT	–	Red
Dolomite (ECE)	1854cc	4	1974/	AUD 603F	F	HS4	–	ABK	–	Red
				AUD 603R	R	HS4	–	ABK	–	Red
Dolomite Sprint	1998cc	4	1973/74	AUD 545F	F	HS6	–	BBT	–	Yellow
				AUD 545R	R	HS6	–	BBT	–	Yellow
Dolomite Sprint (ECE)	1998cc	4	1974	AUD 661F	F	HS6	–	BCM	–	Yellow
				AUD 661R	R	HS6	–	BCM	–	Yellow
2000	1998cc	6	1974/	AUD 604F	F	HS4	–	ABL	–	Yellow
				AUD 604R	R	HS4	–	ABL	–	Yellow
2500 TC	2498cc	6	1974/	AUD 607F	F	HS4	–	ABU	–	Yellow
				AUD 607R	R	HS4	–	ABU	–	Yellow
Stag V8	2997cc	8	1970/74	–	–	–	–	–	–	–

Model Details	Capacity	No. of Cyl.	Year	Spec.	Position	Type	Rich	Needle Std.	Weak	Piston Spring
UNIVERSAL POWER DRIVES										
Unipower	998cc	4	–	AUD 104L	LH	HS2	M	GY	GG	Blue
				AUD 104R	RH	HS2	M	GY	GG	Blue
VANDEN PLAS										
Princess 4 litre DM4	–	6	1956/64	–	–	–	–	–	–	
Princess 3 litre HC & LC	2912cc	6	1961/64	AUD 44F	F	H4	3	AR	HA	Yellow
				AUD 44R	R	H4	3	AR	HA	Yellow
Princess 4 litre R	3909cc	6	1964	AUD 97F	F	HS8	–	UV	–	Red/Blue
				AUD 97R	R	HS8	–	UV	–	Red/Blue
Princess 4 litre R	3909cc	6	1965/66	AUD 215F	F	HS8	–	UV	–	Red/Blue
				AUD 215R	R	HS8	–	UV	–	Red/Blue
Princess 4 litre R (Service replacement)	3909cc	6	1964/66	AUD 418F	F	HS8	–	UV	–	Red/Blue
				AUD 418R	R	HS8	–	UV	–	Red/Blue
Princess 1100	1098cc	4	1964	AUD 69L	LH	HS2	D6	D3	GV	Blue
				AUD 69R	RH	HS2	D6	D3	GV	Blue
Princess 1300	1275cc	4	1967/68	AUD 186	–	HS4	BQ	DZ	CF	Red
Princess Auto	1275cc	4	1967/68	AUD 271	–	HS4	BQ	DZ	CF	Red
Princess 1300	1275cc	4	1968/69	AUD 318L	LH	HS2	M	EB	GG	Blue
				AUD 318R	RH	HS2	M	EB	GG	Blue
Princess 1300	1275cc	4	1969/71	AUD 344L	LH	HS2	M	GY	GG	Blue
				AUD 344R	RH	HS2	M	GY	GG	Blue
Princess 1300	1275cc	4	1971	AUD 431L	LH	HS2	M	GY	GG	Blue
				AUD 431R	RH	HS2	M	GY	GG	Blue
Princess 1300 (ECE)	1275cc	4	1971/72	AUD 454L	LH	HS2	–	AAP	–	Blue
				AUD 454R	RH	HS2	–	AAP	–	Blue
Princess 1300 (ECE)	1275cc	4	1971/72	AUD 496L	LH	HS2	–	AAP	–	Blue
				AUD 496R	RH	HS2	–	AAP	–	Blue
Princess 1500	1485cc	4	1974/	AUD 628	–	HS6	–	BAS	–	Red
VOLVO										
B18B Snow Weasel (Pancake filter)	1788cc	4	1965/66	AUD 95F	F	HS6	–	ZH	–	Red
				AUD 95R	R	HS6	–	ZH	–	Red
B18B Snow Weasel	1788cc	4	1967	AUD 277F	F	HS6	–	ZH	–	Red
				AUD 277R	R	HS6	–	ZH	–	Red
B18B P1800	1788cc	4	1963/65	AUD 94F	F	HS6	–	ZH	–	Red
				AUD 94R	R	HS6	–	ZH	–	Red
B18B 1800S (Pancake filter)	1788cc	4	1965/66	AUD 93F	F	HS6	–	KD	–	Red
				AUD 93R	R	HS6	–	KD	–	Red
B18B 1800S (Silencer, Paper element)	1788cc	4	1965/66	AUD 204F	F	HS6	–	KF	–	Red
				AUD 204R	R	HS6	–	KF	–	Red
B18B 144 (Pancake filter)	1788cc	4	1967/68	AUD 230F	F	HS6	–	KD	–	Red
				AUD 230R	R	HS6	–	KD	–	Red
B18B 144 (Silencer filter)	1788cc	4	1967/68	AUD 231F	F	HS6	–	KF	–	Red
				AUD 231R	R	HS6	–	KF	–	Red
B18B 144	1788cc	4	1968	AUD 305F	F	HS6	–	KN	–	Red
				AUD 305R	R	HS6	–	KN	–	Red
B18B 144 (USA)	1788cc	4	1967/68	AUD 252F	F	HS6	–	DX	–	Red
				AUD 252R	R	HS6	–	DX	–	Red
B18B 144 (USA)	1788cc	4	1968	AUD 331F	F	HS6	–	KN	–	Red
				AUD 331R	R	HS6	–	KN	–	Red
B18D P544 and P122S (Pancake filter)	1788cc	4	1965/66	AUD 94F	F	HS6	–	ZH	–	Red
				AUD 94R	R	HS6	–	ZH	–	Red
B18D P544 and P122S (Oil bath filter)	1788cc	4	1965/66	AUD 200F	F	HS6	–	KE	–	Red
				AUD 200R	R	HS6	–	KE	–	Red
B18D (Silencer filter)	1788cc	4	1966/67	AUD 202F	F	HS6	–	KG	–	Red
				AUD 202R	R	HS6	–	KG	–	Red
B18D 144 (Pancake filter)	1788cc	4	1967/68	AUD 232F	F	HS6	–	SM	–	Red
				AUD 232R	R	HS6	–	SM	–	Red
B20A 142/144	1990cc	4	1969/70	AUD 403	–	HS6	–	BAH	–	Green
B20B 144S	1990cc	4	1969/70	AUD 331F	F	HS6	–	KN	–	Red
				AUD 331R	R	HS6	–	KN	–	Red
B20B 144 (USA)	1990cc	4	1971	AUD 388F	F	HIF6	–	BAL	–	Red
				AUD 388R	R	HIF6	–	BAL	–	Red
B20B 144 (LHD)	1990cc	4	1971/72	AUD 499F	F	HIF6	–	BBB	–	Red
				AUD 499R	R	HIF6	–	BBB	–	Red
B20B 144 Auto (LHD)	1990cc	4	1971/72	AUD 511F	F	HIF6	–	BBB	–	Red
				AUD 511R	R	HIF6	–	BBB	–	Red
B20D 144 (LHD)	1990cc	4	1971	AUD 433F	F	HIF6	–	BAL	–	Red
				AUD 433R	R	HIF6	–	BAL	–	Red
B20D 144 (LHD)	1990cc	4	1972	AUD 522F	F	HIF6	–	BBZ	–	Red
				AUD 522R	R	HIF6	–	BBZ	–	Red
B20B (LHD)	1990cc	4	1972/73	AUD 599F	F	HIF6	–	BBB	–	Red
				AUD 599R	R	HIF6	–	BBB	–	Red
B20B 144 Auto (LHD)	1990cc	4	1972/73	AUD 600F	F	HIF6	–	BBB	–	Red
				AUD 600R	R	HIF6	–	BBB	–	Red
B20B 144 (Canada)	1990cc	4	1973/74	AUD 666F	F	HIF6	–	BBB	–	Red
				AUD 655R	R	HIF6	–	BBB	–	Red
B20B 144 Auto (Canada)	1990cc	4	1973/74	AUD 677F	F	HIF6	–	BBB	–	Red
				AUD 655R	R	HIF6	–	BBB	–	Red
B20A 144 (LHD)	1990cc	4	1974	AUD 466	–	HIF6	–	BCJ	–	Green
B20A 144 (LHD)	1990cc	4	1974/	AUD 699	–	HIF6	–	BCJ	–	Green

Model Details	Capacity	No. of Cyl.	Year	Spec.	Position	Type	Rich	Needle Std.	Weak	Piston Spring
WOLSELEY										
Wolseley 1500	1489cc	4	1962/64	AUC 979	–	HS2	M	GY	GG	Red
Hornet Mk I & Mk II	998cc	4	1963/68	AUD 86	–	HS2	M	GX	GG	Red
Hornet Mk III	998cc	4	1968/69	AUD 298	–	HS2	M	GX	GG	Red
1100	1098cc	4	1965/66	AUD 69L	LH	HS2	D6	D3	GV	Blue
				AUD 69R	RH	HS2	D6	D3	GV	Blue
6/110 (hc & lc)	2912cc	6	1961/64	AUD 43F	F	H4	3	AR	HA	Yellow
				AUD 43R	R	H4	3	AR	HA	Yellow
6/110	2912cc	6	1967	AUD 240F	F	H4	3	AR	HA	Yellow
				AUD 240R	R	H4	3	AR	HA	Yellow
1300	1275cc	4	1967/68	AUD 186	–	HS4	BQ	DZ	CF	Red
1300 Auto	1275cc	4	1967/68	AUD 271	–	HS4	BQ	DZ	CF	Red
1300	1275cc	4	1968/69	AUD 318L	LH	HS2	M	EB	GG	Blue
				AUD 318R	RH	HS2	M	EB	GG	Blue
1300 Mk II	1275cc	4	1969/71	AUD 344L	LH	HS2	M	GY	GG	Blue
				AUD 344R	RH	HS2	M	GY	GG	Blue
1300 Mk II	1275cc	4	1971/74	AUD 431L	LH	HS2	M	GY	GG	Blue
				AUD 431R	RH	HS2	M	GY	GG	Blue
1300 Mk II	1275cc	4	1971/74	AUD 454L	LH	HS2	–	AAP	–	Blue
				AUD 454R	RH	HS2	–	AAP	–	Blue
1300 Mk II	1275cc	4	1971/74	AUD 496L	LH	HS2	–	AAP	–	Blue
				AUD 496R	RH	HS2	–	AAP	–	Blue
18/85 Auto	1798cc	4	1967	AUD 273	–	HS6	SW	TW	CIW	Yellow
18/85 Mk II Auto	1798cc	4	1969/71	AUD 291	–	HS6	SA	ZH	CIW	Yellow
18/85 Mk II S	1798cc	4	1969/71	AUD 171L	LH	HS6	CI	TZ	CIW	Red
				AUD 171R	RH	HS6	CI	TZ	CIW	Red
Wolseley Six	2227cc	6	1972/74	AUD 409F	LH	HS6	–	BBD	–	Red
				AUD 409R	RH	HS6	–	BBD	–	Red
Wolseley Six (ECE)	2227cc	6	1972/	AUD 546F	LH	HIF6	–	BBN	–	Red
				AUD 546R	RH	HIF6	–	BBN	–	Red
Wolseley Six Auto (ECE)	2227cc	6	1972/	AUD 581F	LH	HIF6	–	BBN	–	Red
				AUD 581R	RH	HIF6	–	BBN	–	Red

Special conversions:

This cannot be a complete listing. It covers the majority of the more popular 'production' conversions with specifications where available.

Model Details	Capacity	No. of Cyl.	Year	Spec.	Position	Type	Rich	Needle Std.	Weak	Piston Spring
Citroen 2.6	–	6	1950/54	AUC 712F	F	H4Th[2]	MME	CP4	CF	Red
				AUC 712R	R	H4	MME	CP4	CF	Red
BMC and BRITISH LEYLAND										
Minor MM and Series II	800cc	4	1948/56	–	F	UBA	M9	EK	MOW	Red
- Derrington					R	UBA	M9	EK	MOW	Red
Oxford MO Series II/III	1489cc	4	1950/57	–	F	H2	CJ	HB	MO	Red
- Derrington					R	H2	CJ	HB	MO	Red
Minor - Power drive	948cc	4	1957	–	F	H1	EB	GG	MOW	Red
and Alexander					R	H1	EB	GG	MOW	Red
MG Elva	1588cc	4	1959/61	AUC 892F	F	H4	–	GS	–	Red
				AUC 892R	R	H4	–	GS	–	Red
Minor 1000 Speedwell	948cc	4	1959/61	AUC 919F	F	H2	–	M8	–	Blue
				AUC 919R	R	H2	–	M8	–	Blue
BMC A Series Turner	948cc	4	1959/61	AUC 911F	F	H1	–	BX1	–	–
				AUC 911R	R	H1	–	BX1	–	–
BMC A Series Turner	948cc	4	1959/61	AUC 927F	F	H2	–	M6	–	Red
				AUC 927R	R	H2	–	M6	–	Red
Mini–WHMB	948cc	4	1959/71	AUD 489	–	H4	–	AO	–	Blue
Sprite	948cc	4	1960	–	F	H4	–	A5	–	Blue
					R	H4	–	A5	–	Blue
Sprite Sebring	948cc	4	1960	AUC 930F	F	H2	–	GX	–	Blue
				AUC 930R	R	H2	–	GX	–	Blue
Sprite	1098cc	4	1960	AUC 989F	F	H4	–	AM	–	Blue
				AUC 989R	R	H4	–	AM	–	Blue
Formula Junior BMC	997cc	4	1960	AUC 951	–	H4	–	AM	–	Blue
				AUC 951	–	H4	–	AM	–	Blue
Mangoletsi Remix	–	–	1961/63	AUD 25F	F	H1	–	M8	–	–
				AUD 25R	R	H1	–	M8	–	–
Mini Cooper Thermo Jets	997cc	4	1961/63	AUD 59LH	LH	H4	–	MME	–	Blue
				AUD 59RH	RH	H4	–	MME	–	Blue
Healey 3000 BN7 Competition 9 : 1 cr	2912cc	6	1961	AUD 19F	F	HD8	–	UH	–	Blue/Black
				AUD 19C	C	HD8	–	UH	–	Blue/Black
				AUD 19R	R	HD8	–	UH	–	Blue/Black
Mini Competition	848cc	4	1962/63	AUD 106LH	LH	H4	–	MME	–	Blue
				AUD 106RH	RH	H4	–	MME	–	Blue
Sprite Speedwell	1098cc	4	1962/63	AUD 103F	F	H4	–	AO	–	Red
				AUD 103R	R	H4	–	AO	–	Red
Mini Cooper S Group II	970cc	4	1964/68	AUD 164LH	LH	H4	–	CP4	–	Blue
				AUD 164RH	RH	H4	–	CP4	–	Blue
Mini Cooper S Group II	1071cc	4	1964/68	AUD 108LH	LH	H4	–	MME	–	Blue
				AUD 108RH	RH	H4	–	MME	–	Blue
Mini Cooper S Group II	1275cc	4	1964/68	AUD 165LH	LH	H4	–	BG	–	Blue
				AUD 165RH	RH	H4	–	BG	–	Blue
Formula III Cooper BMC	997cc	4	1964/68	AUD 143	–	HS6	–	UVP	–	Red
Morris (Downton) 1100	1098cc	4	1964/68	AUD 137L	LH	H4	–	AM	–	Blue
				AUD 137R	RH	H4	–	AM	–	Blue
BLMC Maxi (Special Tuning)	1485cc	4	1969/71	AUD 438L	LH	HS4	–	AAB	–	Red
				AUD 438R	RH	HS4	–	AAB	–	Red
FORD										
E93A	1172cc	4	1949/53	–	F	HV1	M9	EK	MOW	–
					R	HV1	M9	EK	MOW	–

Model Details	Capacity	No. of Cyl.	Year	Spec.	Position	Type	Rich	Needle Std.	Weak	Piston Spring
E93A - Dellow	1172cc	4	1950	–	–	HV3	–	RLS	–	Red
30 hp V8 (special adapter)	3622cc	8	1950	–	F	H4	RO	6	–	Red
					R	H4	RO	6	–	Red
Consul (Series 1)	1508cc	4	1952	–	F	H4	–	61	–	Yellow
					R	H4	–	61	–	Yellow
Consul (Series 1)	1508cc	4	1953	–	F	H2	–	62	–	Yellow
					R	H2	–	62	–	Yellow
Zephyr (Series 1)	2262cc	6	1953	–	F	H2	–	WX	–	Yellow
					C	H2	–	WX	–	Yellow
					R	H2	–	WX	–	Yellow
Consul (Series 1) - Dellow	1508cc	4	1953	–	F	H2	–	M5	–	Red
					R	H2	–	M5	–	Red
100E Aquasport	1172cc	4	1953/57	–	F	MC2	M1	A5	HA	Red
					R	MC2	M1	A5	HA	Red
100E Prefect & Anglia	1172cc	4	1953	–	F	H1	–	M6	–	–
					R	H1	–	M6	–	–
100E Lotus	1172cc	4	1954/60	–	F	H2	M5	M6	M7	Red
					R	H2	M5	M6	M7	Red
Consul - Aquaplane (Series 1)	1508cc	4	1954/57	–	F	H4	4	3	L	Red
					R	H4	4	3	L	Red
Zephyr - Aquaplane (Series 1)	2262cc	6	1954/57	–	F	H4	4	3	L	Yellow
					C	H4	4	3	L	Red
					R	H4	4	3	L	Yellow
Zephyr - Raymond Mays (Series 1)	2262cc	6	1954/56	–	F	H4	CN	5	GE	Yellow
					R	H4	CN	5	GE	Yellow
Zephyr (Series 1) - Dellow	2262cc	6	1954	–	F	H2	–	M5	–	Red
					C	H2	–	M5	–	Red
					R	H2	–	M5	–	Red
100E Prefect & Anglia - Dellow	1172cc	4	1955	–	F	H1	M9	EK	MOW	–
					R	H1	M9	EK	MOW	–
Consul WHMB	1508cc	4	1955/57	–	F	H2	H2	QA	QW	Red
					R	H2	H2	QA	QW	Red
Zephyr WHMB	2262cc	6	1955/57	–	F	H2	EM	ES	AP	Red
					C	H2	EM	ES	AP	Red
					R	H2	EM	ES	AP	Red
Consul R. Owen (Series 2) 4 port head	1703cc	4	1958/60	–	–	H6	–	RB	–	Red
Consul R. Owen (Series 2) 6 port head	1703cc	4	1958/60	–	F	H6	–	RB	–	Red
					R	H6	–	RB	–	Red
105E FJ	997cc	4	1960/62	–	F	H4	–	AM	–	Blue
					R	H4	–	AM	–	Blue
100E Aquaplane	1172cc	4	1960/62	–	F	H2	–	GX	–	Blue
					R	H2	–	GX	–	Blue
105E/107E Aquaplane	997cc 1198cc	4	1960/62	–	F	H2	–	A5	–	Blue
					R	H2	–	A5	–	Blue
Consul R. Owen (Series 2) 4 port head	1703cc	4	1962	–	–	H6	–	RB	–	Red
Zephyr Raymond Mays	2553cc	6	1962	–	F	H4	–	AY	–	Yellow
					R	H4	–	AY	–	Yellow
Lotus Ford 105E	997cc	4	1961/62	AUD 38F	F	H2	–	A5	–	Blue
				AUD 38R	R	H2	–	A5	–	Blue
Turner/Classic	1297cc	4	1961/62	AUD 57	–	HS4	–	DJ	–	Red
Formula 3 Holbay - Ford	997cc	4	1964	–	–	HS6	–	UVP	–	Red
Escort 1100 & 1300	1098cc 1298cc	4	1968/	AUD 674	–	HS2	–	AAC	–	Red
HILLMAN										
Minx	1390cc	4	1956/58	AUC 825F	F	H2	CU	CZ	CF	Blue
				AUC 825R	R	H2	CU	CZ	CF	Blue
Alexander Minx	1600cc	4	1959/61	AUC 923F	F	H2	–	GR	–	Blue
				AUC 923R	R	H2	–	GR	–	Blue
Minx	1600cc	4	1964	AUD 145F	F	H4	–	QA	–	Red
				AUD 145R	R	H4	–	QA	–	Red
Imp	875cc	4	1964	AUD 140F	F	HS2	–	H4	–	Blue
				AUD 140R	R	HS2	–	H4	–	Blue
JAGUAR										
E Type V12	5343cc	12	1972/	AUD 547 NSF	NS F	HIF6	–	BBG	–	Yellow
				AUD 547 NSR	NS R	HIF6	–	BBG	–	Yellow
				AUD 547 OSF	OS F	HIF6	–	BBG	–	Yellow
				AUD 547 OSR	OS R	HIF6	–	BBG	–	Yellow
TRIUMPH										
Alexander Herald	948cc	4	1960/61	–	F	H2	–	M6	–	Blue
					R	H2	–	M6	–	Blue
Spitfire Group II	1147cc	4	1966	AUD 235F	F	H4	–	DB	–	Blue
				AUD 235R	R	H4	–	DB	–	Blue
Vitesse	1596cc	6	1963/64	AUD 98F	F	HS2	–	MO	–	Red
				AUD 98R	R	HS2	–	MO	–	Red
2000	1998cc	6	1966/73	AUD 704F	F	HS4C	–	ABL	–	Yellow
				AUD 704R	R	HS4C	–	ABL	–	Yellow
Vitesse	1998cc	6	1966/71	AUD 704F	F	HS4C	–	ABL	–	Yellow
				AUD 704R	R	HS4C	–	ABL	–	Yellow

Appendix 2 - Needle charts

Notes on needles

Needle selection

The applications list (Appendix 1) usually details three alternative needles – Rich, Standard, Weak – for each application of carburettor to a particular car. These recommendations apply only to the installation as designed by the relevant vehicle manufacturer.

The standard needle should normally be used except in special circumstances.

The rich needle gives a richer mixture ratio, and therefore greater power, throughout its range.

The weak needles for carburettor specifications fitted with fixed needles are not economy needles – they are to compensate for operation at high altitudes. There is no guarantee that greater economy will result from their use at normal altitudes.

.090 Jet Needles

	A5	AA	AB	AC	AC2	AD	AE
1	.089	.089	.089	.089	.089	.089	.089
2	.085	.085	.085	.085	.085	.085	.085
3	.0826	.080	.080	.082	.082	.082	.081
4	.080	.0767	.0785	.080	.080	.080	.078
5	.0782	.0735	.0768	.0783	.0783	.0780	.0763
6	.0765	.071	.075	.0765	.0765	.0760	.0754
7	.0746	.0689	.0732	.0746	.0746	.0740	.0745
8	.073	.0661	.0718	.073	.073	.0720	.0737
9	.0711	.0638	.0702	.071	.071	.070	.0728
10	.0694	.0614	.0688	.0694	.0694	.0680	.0718
11	.0676	.0591	.0671	.0676	.0676	.066	.071
12	.066	.0566	.0657	.066	.066	.064	.070
13		.054	.064	.064	.064	.062	

	AF	AG	AH	AH1	AH2	AI	AJ
1	.089	.089	.089	.089	.089	.089	.089
2	.085	.085	.0862	.086	.085	.085	.085
3	.0814	.0795	.083	.082	.082	.0817	.0815
4	.078	.0745	.0803	.079	.0794	.0798	.079
5	.0758	.0702	.0775	.0765	.077	.078	.0767
6	.0727	.0665	.0756	.075	.0748	.0765	.0745
7	.071	.063	.0733	.073	.0726	.075	.0723
8	.0695	.0598	.0711	.071	.0704	.0732	.0703
9	.068	.0567	.069	.069	.0683	.0712	.0683
10	.0665	.054	.067	.067	.0662	.0693	.0663
11	.065	.051	.065	.065	.064	.0685	.064
12	.0632	.0485	.063	.063	.062	.0675	.062
13		.046	.061	.061	.060		

	AK	AL	AM	AN	AO	AP	AQ
1	.089	.089	.089	.089	.089	.089	.089
2	.086	.085	.085	.0855	.085	.085	.085
3	.0825	.0816	.081	.0827	.082	.0817	.080
4	.0795	.0796	.078	.0807	.0793	.0796	.076
5	.0786	.0781	.0753	.0787	.0766	.0777	.0724
6	.078	.077	.073	.077	.0737	.0765	.0694
7	.077	.076	.0704	.0753	.0705	.0752	.0668
8	.0764	.0748	.068	.074	.0673	.0745	.0642
9	.0755	.0738	.0655	.073	.064	.0736	.062
10	.0747	.0726	.063	.072	.0608	.0727	.060
11	.0738	.0715	.0606	.071	.0576	.072	.058
12	.073	.0705	.0583	.070	.0544	.071	.0558
13			.056	.069	.051		.0536

	AR	AS	AT	AU	AV	AW	AX
1	.089	.089	.088	.089	.089	.090	.089
2	.085	.0845	.0856	.084	.085	.085	.0843
3	.082	.079	.0833	.0815	.0805	.0807	.0807
4	.0795	.075	.0809	.079	.0773	.078	.0775
5	.0771	.072	.0785	.0773	.0742	.0757	.075
6	.0748	.0692	.0761	.0755	.0717	.0735	.073
7	.073	.0665	.0738	.0737	.070	.0713	.071
8	.0712	.0635	.0714	.0717	.0675	.0693	.0692
9	.0696	.061	.069	.0698	.065	.0674	.0675
10	.068	.0586	.0666	.068	.0625	.0655	.066
11	.066	.056	.0643	.066	.060	.0637	.0645
12	.064	.0533	.0619	.064	.059	.0618	.063
13		.051			.058	.060	.0615

	AY	AZ	† AAA	* AAB	* AAC	* AAD	* AAE
1	.089	.089	.089	.089	.089	.089	.089
2	.085	.085	.085	.085	.0855	.0855	.085
3	.0805	.0815	.0814	.0824	.0835	.0827	.0833
4	.0768	.079	.0785	.0785	.0811	.0807	.0803
5	.0741	.0755	.0755	.0752	.0788	.0787	.0773
6	.072	.071	.072	.0715	.0765	.077	.0745
7	.0694	.0662	.0674	.067	.0742	.0753	.0715
8	.0669	.0615	.063	.061	.072	.074	.068
9	.0643	.0575	.060	.056	.0698	.073	.0653
10	.0617	.0532	.058	.0537	.0676	.072	.0627
11	.059	.0490	.056	.0516	.0655	.071	.060
12	.0565	.0445	.0540	.0493	.0631	.070	.059
13	.0538	.0405	.052	.047	.061	.069	.058
14			.050	.0448	.059	.068	.057
15			.048	.0448	.057	.067	.056
16			.046	.0448	.055	.066	.055

	* AAF	* AAG	* AAH	* AAJ	* AAK	* AAL	* AAM
1	.089	.089	.089	.089	.089	.089	.089
2	.0855	.085	.0855	.0855	.0855	.085	.085
3	.0827	.083	.0832	.0832	.0830	.83	.0824
4	.080	.080	.0812	.0812	.0807	.080	.0785
5	.0775	.0773	.0790	.0789	.0784	.0771	.0752
6	.075	.0745	.0764	.0770	.0760	.0745	.0715
7	.0715	.0715	.0736	.0747	.0735	.0715	.0673
8	.068	.0686	.0705	.0716	.0708	.068	.063
9	.0653	.0658	.0674	.0672	.0677	,0653	.059
10	.0627	.0647	.0644	.0640	.0645	.0627	.0567
11	.060	.0636	.0619	.0620	.0613	.060	.0543
12	.059	.0625	.0593	.0600	.0580	.059	.0519
13	.058	.0614	.0566	.0580	.0550	.058	.0496
14	.057	.0605	.0540	.0560	.0520	.057	.0472
15	.056	.0595	.0515	.0540	.0490	.056	.0448
16	.055	.0584	.0490	.0520	.0460	.055	.0448

	* AAN	* AAP	* AAQ	* AAR	* AAS	* AAT	* AAU
1	.089	.089	.089	.089	.089	.089	.089
2	.0855	.0855	.0855	.085	.085	.0855	.085
3	.0835	.0835	.0827	.0827	.083	.0832	.0828
4	.0815	.0811	.0801	.0795	.0802	.081	.0806
5	.0797	.0788	.0775	.077	.0775	.0784	.078

	* AAN	* AAP	* AAQ	* AAR	* AAS	* AAT	* AAU
6	.0781	.077	.0754	.0745	.075	.076	.0745
7	.0767	.076	.0732	.0715	.0718	.0745	.071
8	.0757	.075	.0718	.0686	.0688	.0725	.0675
9	.0750	.074	.0701	.0658	.0665	.0708	.0647
10	.0745	.073	.0688	.0647	.0648	.0691	.062
11	.074	.072	.0672	.0636	.063	.0674	.059
12	.0735	.071	.0655	.0625	.0612	.0657	.056
13	.073	.070	.0642	.0614	.0596	.064	.053
14	.0725	.069	.0625	.0605	.0577	.0623	.050
15	.072	.068	.0612	.0595	.056	.0606	.047
16	.0715	.067	.0595	.0584	.0542	.0589	.044

Needles marked thus * are supplied with collar attached to shank for use as spring-loaded needles.

† The AAA needle is available without collar as a standard .090 needle and with collar for spring-loaded application.

	* AAV	* AAW	* AAX	* AAY	* AAZ	ABA	ABB
1	.089	.089	.089	.089	.089	.089	.089
2	.0855	.0855	.0855	.0855	.0855	.0855	.0855
3	.0835	.0832	.0833	.083	.0832	.0825	.083
4	.0815	.081	.081	.0807	.0805	.0802	.0807
5	.0792	.0788	.079	.0782	.0777	.0775	.0782
6	.0768	.0765	.0772	.0758	.0754	.075	.0758
7	.0746	.0742	.0758	.0744	.0732	.0718	.0745
8	.072	.0715	.075	.0732	.0718	.0688	.0725
9	.0698	.0683	.0745	.0723	.0701	.0665	.0708
10	.0674	.0652	.074	.0717	.0688	.0648	.0691
11	.065	.062	.0735	.0712	.0672	.063	.0674
12	.0627	.059	.073	.0707	.0655	.0612	.0657
13	.0602	.056	.0725	.0700	.0642	.0596	.064
14	.0578	.053	.072	.069	.0625	.0577	.0623
15	.0554	.050	.0715	.068	.0612	.056	.0606
16	.053	.047	.071	.067	.0595	.0542	.0589

	* ABC	* ABD	* ABE	* ABF	* ABG	* ABH	* ABJ
1	.089	.089	.089	.089	.089	.089	.089
2	.0855	.085	.085	.0855	.0855	.085	.085
3	.0827	.0828	.083	.0832	.0832	.0825	.083
4	.0810	.0803	.080	.0809	.0807	.079	.080
5	.077	.0765	.0775	.0785	.0784	.076	.0765
6	.076	.074	.0758	.0761	.0760	.073	.0725
7	.0745	.071	.0717	.0737	.0737	.0705	.0715
8	.0725	.0675	.0699	.0712	.0717	.0693	.0686
9	.0708	.0647	.0676	.0687	.0697	.0682	.0658
10	.0691	.062	.0653	.0664	.0677	.067	.0647
11	.0674	.059	.063	.0642	.0656	.0656	.0636
12	.0657	.056	.0612	.0618	.0635	.064	.0625
13	.064	.053	.0596	.0595	.0614	.063	.0614
14	.0623	.050	.0577	.0572	.0593	.062	.0605
15	.0606	.047	.056	.0550	.0572	.061	.0595
16	.0589	.044	.0542	.0525	.0500	.060	.0584

	* ABK	* ABL	* ABM	* ABN	* ABP	* ABQ	* ABR
1	.089	.089	.089	.089	.089	.089	.089
2	.0855	.0855	.0855	.0855	.0855	.0855	.0855
3	.0833	.0832	.0835	.0832	.0831	.0828	.0830
4	.0812	.0810	.0815	.0805	.0808	.0803	.0803
5	.0793	.0790	.0795	.0777	.0787	.0778	.0775
6	.0775	.0770	.0777	.0754	.0766	.0753	.0745
7	.0757	.0757	.0762	.0732	.0743	.0730	.0715
8	.0742	.0745	.075	.0712	.0722	.0707	.0681
9	.0728	.0737	.074	.0692	.0700	.0685	.0653
10	.0717	.0730	.073	.0672	.0685	.0660	.0620
11	.0704	.0723	.072	.0652	.0672	.0638	.0600
12	.0692	.0715	.071	.0632	.0659	.0615	.0590
13	.0680	.0709	.070	.0612	.0646	.0590	.0580
14	.0668	.0700	.069	.0592	.0633	.0570	.0570
15	.0656	.0694	.068	.0572	.062	.0550	.0560
16	.0644	.0686	.067	.0552	.0607	.0530	.0550

	* ABS	* ABV	* ABW	* ABY	* ABZ	BA	BB
1	.089	.089	.089	.089	.089	.089	.089
2	.0855	.0855	.0865	.085	.0865	.0856	.085
3	.0835	.0835	.0840	.0827	.0840	.0822	.0825
4	.0814	.0815	.0815	.0798	.0815	.0805	.080
5	.0789	.0797	.0786	.0771	.0790	.0794	.0775
6	.0765	.0775	.0761	.0720	.0761	.0777	.075
7	.0746	.0732	.0728	.0653	.0730	.0760	.0725
8	.0728	.0720	.0704	.0621	.0704	.0743	.070
9	.0713	.0708	.0680	.0589	.0676	.0727	.0675
10	.0692	.0691	.0663	.0557	.0650	.0710	.065
11	.0674	.0674	.0637	.0525	.0624	.0694	.0625
12	.0656	.0657	.0614	.0493	.0596	.0677	.060
13	.064	.0640	.0590	.0461	.0570		.0575
14	.062	.0623	.0567	.0429	.0544		
15	.060	.0606	.0543	.0429	.0517		
16	.058	.0589	.0519	.0429	.0490		

	BD	BE	BF	BG	BH	BI	BJ
1	.090	.089	.089	.089	.089	.089	.089
2	.0856	.084	.085	.085	.085	.0855	.0855
3	.0822	.0805	.082	.0815	.0812	.0808	.0824
4	.0805	.0773	.0796	.0782	.0775	.0777	.0794
5	.0794	.074	.0764	.0745	.0753	.0751	.0769
6	.0777	.0705	.072	.0695	.074	.073	.0749
7	.0760	.067	.068	.0647	.0731	.0714	.0734
8	.0750	.0634	.0635	.060	.0722	.0705	.0725
9	.0740	.060	.0591	.0557	.0714	.0701	.0721
10	.0730	.0565	.0549	.0515	.0705	.0697	.0717
11	.0720	.053	.0505	.0474	.0696	.0694	.0714
12	.0710	.0495	.0463	.043	.0688	.069	.071
13		.046	.042	.039	.068		

	BK	BL	BM	BN	BO	BP	BQ
1	.089	.089	.089	.089	.089	.089	.089
2	.0855	.0855	.0855	.0855	.0855	.085	.085
3	.0815	.081	.0805	.0816	.0835	.0814	.082
4	.0785	.0777	.0768	.0784	.0815	.078	.079
5	.0762	.075	.074	.0758	.080	.074	.075
6	.0738	.073	.072	.0738	.0787	.0715	.0725
7	.0722	.0715	.0705	.0724	.077	.0695	.0705
8	.0715	.0707	.0695	.0714	.0757	.0673	.0685
9	.071	.070	.0692	.0707	.075	.065	.0662
10	.0706	.0698	.0688	.0703	.074	.0625	.064
11	.0703	.0693	.0684	.0701	.073	.060	.0616
12	.070	.069	.068	.070	.072	.058	.0594
13	.0695	.0688			.071	.056	.067

	BR	BS	BT	BU	BV	BW	BX
1	.089	.088	.088	.090	.089	.089	.089
2	.085	.0856	.0856	.085	.0845	.0855	.085
3	.0817	.0836	.0835	.080	.081	.0818	.0805
4	.0785	.0817	.0813	.077	.0785	.079	.077
5	.0757	.0798	.0792	.0745	.076	.0765	.075
6	.075	.0778	.0771	.072	.074	.0742	.0735
7	.0741	.076	.0749	.0695	.072	.0718	.072
8	.0735	.074	.0726	.0675	.070	.0693	.0705
9	.0728	.072	.0705	.0655	.068	.067	.0687
10	.072	.0701	.0684	.0625	.066	.0645	.067
11	.071	.0683	.0663	.0602	.064	.062	.0655
12	.070	.0664	.064	.058	.062	.0596	.064
13	.069			.056	.060	.057	

	BX1	BY	BZ	BB2	BB3	BB4	C
1	.089	.088	.088	.0890	.0890	.0890	.089
2	.085	.084	.0843	.085	.085	.085	.0845
3	.0827	.0805	.081	.081	.081	.081	.081
4	.081	.0775	.0752	.0787	.078	.078	.077
5	.0792	.0745	.0719	.0762	.075	.0742	.0742
6	.0777	.0725	.0686	.0735	.0718	.0706	.071
7	.076	.0709	.0653	.0712	.0687	.067	.0683
8	.075	.069	.0622	.0685	.0654	.0638	.066
9	.074	.067	.059	.0659	.0622	.0602	.0635
10	.073	.0653	.056	.0632	.0589	.0564	.0613
11	.072	.0635	.053	.0604	.0554	.0528	.0594
12	.071	.062	.050	.0577	.0526	.0495	.0575
13		.060	.047	.055	.0494	.046	.056

	CA	CB	CC	CD	CE	CF	CG
1	.089	.089	.089	.089	.089	.089	.090
2	.0855	.085	.085	.085	.085	.085	.084
3	.084	.0805	.081	.0815	.082	.083	.081
4	.0825	.077	.078	.0775	.0795	.0805	.079
5	.081	.0745	.075	.0735	.0777	.0787	.0765
6	.0807	.0725	.0725	.070	.076	.0767	.075

	CA	CB	CC	CD	CE	CF	CG
7	.0803	.071	.0698	.069	.0745	.0747	.0735
8	.0803	.0695	.067	.0678	.0735	.0727	.0722
9	.0803	.068	.064	.0666	.072	.0707	.0707
10	.0803	.0665	.061	.0654	.071	.0687	.0693
11	.0803	.065	.058	.0643	.0695	.0667	.0678
12	.0803	.0635	.055	.063	.0682	.0647	.0664
13	.0803	.062	.052	.062	.067		.065

	CH	CJ	CK	CL	CM	CN	CO
1	.090	.090	.090	.090	.088	.089	.089
2	.084	.084	.084	.084	.084	.085	.085
3	.0805	.0815	.081	.080	.0805	.0812	.081
4	.0782	.0795	.079	.0775	.0775	.078	.0787
5	.0767	.0775	.0775	.076	.074	.075	.0765
6	.076	.0761	.077	.075	.0718	.072	.0747
7	.0756	.0747	.0766	.0746	.070	.069	.073
8	.0752	.0734	.0762	.0742	.0685	.0665	.0725
9	.0747	.072	.0757	.0737	.0668	.064	.072
10	.0743	.0705	.0753	.0733	.0652	.062	.0717
11	.0738	.0692	.0749	.0728	.0635	.060	.0714
12	.0733	.0677	.0745	.0724	.062	.058	.071
13	.073	.0664	.074	.072	.060	.056	

	CP	CP4	CQ	CR	CS	CS1	CS2
1	.089	.088	.088	.088	.089	.089	.089
2	.085	.0852	.0852	.0852	.085	.085	.085
3	.0813	.0825	.0825	.083	.0822	.081	.081
4	.0793	.079	.0798	.0805	.0792	.077	.077
5	.0775	.0757	.0768	.078	.0765	.073	.073
6	.0757	.0725	.0737	.0754	.0725	.0692	.069
7	.074	.069	.0706	.0725	.0706	.066	.065
8	.0735	.0655	.0676	.0697	.069	.063	.0615
9	.073	.062	.0646	.067	.0672	.060	.058
10	.0725	.0585	.0615	.064	.0655	.057	.0546
11	.072	.0545	.0585	.0613	.0638	.054	.051
12	.0715	.051	.0555	.0585	.062	.051	.0475
13		.047	.0525	.0556	.0605	.048	.044

	CT	CU	CW	CX	CY	CZ	D1
1	.089	.089	.089	.089	.089	.089	.089
2	.085	.085	.085	.085	.085	.085	.085
3	.081	.0825	.081	.081	.080	.0827	.082
4	.077	.080	.079	.0796	.076	.0806	.0802
5	.0738	.0775	.078	.0788	.0738	.0785	.0793
6	.0706	.0735	.077	.078	.0715	.0745	.0785
7	.0675	.0715	.0763	.0771	.0695	.0727	.0776
8	.0643	.070	.0754	.0763	.068	.071	.077
9	.061	.0683	.0745	.0755	.067	.0693	.0764
10	.058	.0665	.0736	.0748	.066	.0675	.0759
11	.0547	.0647	.0727	.074	.0655	.0657	.0752
12	.0515	.0630	.0718	.073	.065	.064	.0748
13	.0485	.0610			.0645	.0625	

	D2	D3	D4	D6	D7	D8	D9
1	.089	.089	.089	.089	.089	.089	.089
2	.085	.085	.0855	.0855	.0855	.085	.0852
3	.0815	.083	.0825	.0825	.0817	.0817	.0825
4	.0800	.081	.0805	.0805	.0795	.0785	.0806
5	.0795	.080	.079	.0793	.0785	.0767	.0791
6	.0790	.0794	.078	.0785	.0778	.076	.0778
7	.0785	.0785	.0767	.0776	.077	.075	.0764
8	.0780	.0778	.0756	.077	.0765	.074	.075
9	.0775	.0772	.0745	.0764	.0759	.073	.0736
10	.0770	.0764	.0733	.0759	.0752	.072	.0723
11	.0765	.0757	.0721	.0752	.0747	.071	.071
12	.0760	.075	.071	.0748	.074	.070	.0696

	DA	DB	DC	DD	DE	DH	DJ
1	.089	.089	.089	.089	.089	.089	.089
2	.084	.085	.084	.0855	.0855	.085	.085
3	.082	.082	.082	.0835	.0825	.0822	.0822
4	.080	.080	.0805	.0817	.0802	.080	.0795
5	.0788	.078	.0796	.0798	.0772	.0775	.0765
6	.0784	.0753	.0794	.0782	.0745	.074	.073
7	.0780	.0717	.0792	.0767	.0734	.072	.071
8	.0776	.0674	.0790	.0752	.0729	.071	.070
9	.0773	.062	.0787	.0740	.0723	.070	.069
10	.0769	.0557	.0785	.0730	.0717	.069	.068
11	.0765	.0493	.0783	.0720	.0712	.068	.067
12	.0761	.043	.0780	.0710	.0707	.067	.066
13		.0368				.066	.065

	DK	DL	DM	DN	DP	DQ	DR
1	.089	.089	.089	.0885	.089	.089	.088
2	.085	.085	.084	.085	.085	.085	.085
3	.0817	.082	.081	.082	.081	.0815	.080
4	.0787	.0795	.0791	.080	.0777	.0785	.0768
5	.0755	.077	.0780	.0782	.075	.076	.0735
6	.072	.0745	.0775	.077	.0735	.073	.0705
7	.070	.0715	.0770	.076	.0723	.071	.0675
8	.069	.070	.0765	.075	.0715	.070	.065
9	.068	.0685	.0760	.074	.071	.069	.0635
10	.067	.067	.0755	.0728	.0703	.068	.0627
11	.066	.0655	.0750	.0717	.0698	.067	.0618
12	.056	.064	.0745	.0705	.069	.066	.0608
13	.064	.0625			.0688	.065	.060

	DS	DT	DU	DV	DW	DY	DZ
1	.089	.088	.089	.089	.089	.089	.089
2	.085	.084	.085	.084	.085	.085	.085
3	.0805	.081	.081	.0822	.0815	.079	.0827
4	.0776	.078	.0775	.081	.0795	.0765	.0795
5	.0759	.076	.0755	.0792	.0780	.074	.077
6	.074	.074	.074	.0781	.0765	.0715	.0745
7	.0725	.0723	.073	.078	.0752	.069	.0715
8	.071	.0712	.072	.078	.074	.0665	.070
9	.0695	.0709	.071	.078	.0728	.0643	.0685
10	.068	.0705	.070	.078	.0714	.062	.067
11	.067	.0703	.069	.078	.070	.061	.0655
12	.066	.0701	.068	.078	.0686	.060	.064
13		.070				.059	.0625

	E2	E21	E3	E31	E4	EA	EB
1	.088	.089	.088	.088	.088	.089	.089
2	.0845	.085	.0862	.0845	.084	.085	.0855
3	.081	.0826	.083	.0817	.080	.081	.0835
4	.0775	.0788	.0803	.079	.078	.078	.0815
5	.075	.0763	.0775	.076	.076	.075	.0795
6	.0725	.0736	.0747	.073	.0739	.072	.0777
7	.070	.0711	.072	.070	.0718	.0695	.0762
8	.0675	.0685	.0693	.0675	.0695	.067	.075
9	.0646	.066	.0665	.0646	.0673	.0645	.074
10	.0626	.064	.0638	.0626	.0631	.062	.073
11	.0605	.062	.062	.0605	.063	.0595	.072
12	.059	.060	.060	.059	.061	.057	.071
13	.057	.058	.0582	.057	.059	.0545	

	EC	ED	EE	EF	EG	EH	EI
1	.089	.089	.089	.089	.088	.089	.089
2	.085	.085	.085	.085	.085	.085	.085
3	.081	.0825	.080	.082	.080	.082	.081
4	.0775	.0805	.0777	.0795	.0768	.0805	.0775
5	.0740	.0785	.075	.077	.074	.0788	.073
6	.0705	.076	.0735	.074	.071	.0775	.069
7	.0675	.074	.072	.072	.0685	.077	.066
8	.0645	.072	.0715	.071	.0665	.077	.0635
9	.0625	.070	.0709	.070	.065	.077	.0618
10	.0605	.068	.0703	.069	.0637	.077	.060
11	.0585	.066	.0696	.068	.062	.077	.059
12	.0570	.064	.069	.067	.061	.077	.058
13	.0555	.062		.066	.060	.077	.057

	EJ	EK	EL	EM	EN	EO	EP
1	.089	.089	.089	.089	.089	.088	.089
2	.085	.085	.085	.085	.085	.085	.085
3	.0805	.0827	.080	.081	.0813	.0806	.0835
4	.076	.081	.0775	.078	.0778	.0773	.0815
5	.0715	.0792	.0747	.0763	.074	.0746	.0795
6	.068	.0777	.072	.0747	.0706	.072	.0775
7	.065	.0762	.070	.073	.068	.0697	.0755
8	.0625	.075	.068	.0725	.0657	.0877	.075
9	.0605	.075	.066	.072	.0643	.0668	.075
10	.059	.075	.064	.0715	.0632	.0659	.075
11	.058	.075	.062	.071	.062	.065	.075
12	.057	.075	.060	.0705	.061	.064	.075
13	.056		.058		.060	.063	

	EQ	ER	ES	ET	EU	EV	EW
1	.089	.089	.089	.088	.089	.090	.089
2	.085	.085	.085	.086	.085	.0855	.085
3	.0817	.0833	.0818	.0845	.0825	.0815	.0825
4	.0785	.081	.079	.0825	.0805	.078	.081
5	.076	.0777	.077	.0803	.0785	.075	.079

	EQ	ER	ES	ET	EU	EV	EW
6	.0746	.074	.0755	.0781	.077	.0725	.0775
7	.0732	.0725	.0748	.0773	.0755	.0705	.0755
8	.072	.0712	.074	.077	.0745	.069	.0745
9	.071	.0706	.073	.077	.074	.068	.074
10	.070	.0706	.073	.077	.074	.067	.074
11	.069	.0706	.073	.077	.074	.066	.074
12	.068		.073			.065	

	EX	EX/1	EY	EZ	FA	FB	FC
1	.088	.088	.088	.088	.089	.089	.090
2	.0862	.0865	.0856	.0865	.085	.085	.0865
3	.0835	.084	.0835	.0838	.081	.0806	.083
4	.081	.0816	.0812	.0798	.077	.0767	.0795
5	.0785	.0794	.0789	.0775	.072	.0729	.076
6	.076	.077	.0741	.0751	.067	.070	.073
7	.0735	.0745	.0716	.0737	.063	.0687	.071
8	.071	.072	.0708	.0726	.060	.0676	.070
9	.0685	.0698	.0704	.0716	.0585	.0667	.069
10	.066	.0675	.0701	.0706	.0576	.0658	.068
11	.0635	.065	.0698	.0696	.0567	.065	.067
12	.061	.0627			.056	.0643	.066
13	.058	.0602			.055	.0635	.065

	FD	FE	FF	FG	FH	FI	FJ
1	.088	.089	.089	.089	.089	.089	.090
2	.0865	.085	.0856	.085	.085	.085	.085
3	.084	.081	.0822	.0795	.082	.0825	.081
4	.0805	.0775	.079	.075	.080	.079	.0775
5	.078	.074	.0757	.071	.079	.076	.076
6	.076	.071	.072	.0678	.078	.073	.0745
7	.075	.0695	.0700	.065	.077	.0705	.0725
8	.074	.0686	.0672	.0625	.076	.0693	.071
9	.073	.0678	.0651	.060	.075	.0682	.070
10	.072	.0671	.0639	.058	.074	.067	.068
11	.071	.0664	.0622	.056	.073	.0656	.0665
12		.0657		.054	.072	.064	.065
13		.065		.052		.063	

	FK	FL	FM	FN	FO	FP	FQ
1	.089	.089	.090	.089	.089	.0898	.0890
2	.085	.085	.083	.085	.085	.0855	.0850
3	.083	.079	.0796	.081	.0818	.0826	.0820
4	.0812	.0765	.0772	.0775	.0785	.0799	.0795
5	.0796	.074	.075	.0735	.076	.0778	.0774
6	.0784	.0715	.0727	.069	.0745	.0762	.0758
7	.0772	.069	.0703	.066	.0731	.0754	.0745
8	.076	.0665	.068	.063	.0729	.0750	.0735
9	.0748	.065	.0657	.061	.0727	.0743	.0725
10	.0736	.065	.0635	.058	.0725	.0735	.0715
11	.0724	.065	.0612	.055	.0723	.0728	.0705
12	.0712		.060		.0721	.0720	.0695

	FR	FS	FT	FU	FV	FW	FX
1	.089	.089	.089	.089	.089	.089	.089
2	.085	.085	.085	.085	.085	.085	.0855
3	.081	.0827	.082	.082	.083	.081	.0827
4	.078	.081	.0795	.080	.080	.0775	.080
5	.075	.0785	.0772	.078	.0773	.075	.0775
6	.0728	.076	.0753	.0763	.0745	.0725	.075
7	.0705	.075	.0738	.075	.0715	.070	.0715
8	.0685	.074	.0728	.074	.0686	.0675	.068
9	.0663	.073	.0724	.0733	.0658	.0665	.0653
10	.0642	.072	.072	.0728	.0647	.0665	.0627
11	.0622	.071	.072	.0724	.0636	.0665	.060
12	.060	.070	.072	.0714	.0625	.0665	.059
13					.0614	.0665	.058

	FY	FZ	G2	GA	GB	GC	GD
1	.090	.089	.0875	.089	.089	.089	.089
2	.085	.085	.0835	.085	.085	.085	.085
3	.081	.0813	.081	.0795	.082	.083	.0833
4	.0775	.0789	.0785	.077	.0795	.080	.081
5	.075	.077	.0765	.074	.0771	.0773	.079
6	.0735	.0756	.0745	.0715	.0748	.0740	.0766
7	.072	.0748	.0725	.069	.073	.0705	.0753
8	.071	.074	.0705	.067	.0712	.0665	.0743
9	.070	.0735	.069	.065	.0696	.0630	.0738
10	.068	.073	.0674	.0632	.0685	.0605	.0738
11	.066	.0725	.066	.0615	.068	.0590	.0738
12	.064	.072	.0642	.0597	.067	.0575	.0738
13	.062			.058		.0560	

	GE	GE/R	GF	GG	GH	GI	GJ
1	.089	.089	.089	.089	.089	.089	.089
2	.0845	.085	.085	.087	.0845	.085	.084
3	.082	.0825	.0795	.0845	.082	.0825	.0817
4	.0793	.0795	.077	.0825	.0786	.079	.0792
5	.0766	.076	.0745	.0808	.0753	.075	.0770
6	.0739	.0725	.073	.079	.0695	.0715	.0745
7	.0712	.069	.072	.078	.0635	.069	.0717
8	.0686	.066	.0715	.077	.0586	.067	.0696
9	.0656	.064	.071	.0758	.0570	.0665	.0675
10	.0643	.062	.0703	.0745	.0565	.0665	.0654
11	.0627	.0615	.0696	.0732	.0565	.0665	.0632
12	.0627	.061	.069	.072	.0565	.0665	.0611
13	.0627	.0605					.059

	GL	GM	GN	GO	GP	GR	GS
1	.089	.089	.089	.089	.089	.089	.089
2	.084	.086	.0855	.086	.085	.085	.085
3	.082	.0842	.0835	.0845	.080	.082	.0815
4	.0795	.0817	.081	.0825	.076	.079	.0785
5	.0772	.0782	.078	.0792	.075	.077	.0755
6	.075	.0770	.0755	.0777	.0725	.075	.0725
7	.0727	.0758	.0725	.0765	.0696	.073	.070
8	.0706	.0750	.070	.0755	.0666	.071	.0675
9	.0685	.0745	.0675	.0750	.0636	.069	.0650
10	.0664	.0740	.0650	.0745	.0606	.067	.0625
11	.0642	.0735	.0625	.0740	.0577	.065	.060
12	.0621	.0730	.060	.0735	.055	.063	.0575
13	.060		.0575	.0730	.052	.061	.055

	GT	GU	GV	GW	GX	GY	GZ
1	.089	.089	.089	.089	.089	.089	.089
2	.085	.0855	.0855	.085	.0855	.0855	.0855
3	.082	.083	.0836	.083	.0835	.0832	.0835
4	.0795	.0805	.082	.080	.0815	.0812	.0811
5	.077	.0775	.0804	.0775	.0795	.079	.0788
6	.0745	.0745	.080	.076	.0775	.077	.0765
7	.071	.0705	.0796	.0745	.0755	.0753	.0742
8	.0675	.067	.0793	.073	.0735	.074	.072
9	.0640	.0635	.0793	.0715	.072	.073	.0698
10	.0605	.060	.0793	.070	.070	.072	.0676
11	.0570	.0565	.0793	.0685	.068	.071	.0655
12	.0535	.053	.0793	.067	.066	.070	.0631
13	.050		.0793	.066	.064	.069	.061

	H1	H2	H4	H6	HA	HB	HC
1	.088	.088	.089	.089	.089	.089	.089
2	.085	.085	.085	.0855	.085	.085	.0855
3	.082	.082	.081	.082	.0825	.0825	.0822
4	.0792	.0792	.0778	.080	.0805	.0805	.0805
5	.0762	.0762	.076	.078	.0785	.0785	.0785
6	.0735	.0735	.0741	.076	.077	.0776	.077
7	.0707	.0707	.072	.074	.0756	.0764	.0755
8	.0686	.0684	.0702	.072	.074	.0752	.0751
9	.0665	.0661	.0683	.070	.0727	.0743	.0747
10	.0644	.0638	.0663	.068	.0711	.0732	.0743
11	.0622	.0615	.064	.066	.0698	.072	.0739
12	.0601	.0592	.062	.064	.068	.071	.0735
13	.058	.057			.067	.070	

	HD	HE	HF	HG	HV2	HV3	HV4
1	.089	.090	.090	.089	.089	.089	.089
2	.0855	.0845	.0845	.0845	.084	.084	.084
3	.0825	.0817	.082	.0823	.080	.080	.080
4	.081	.080	.0805	.0802	.0778	.078	.0782
5	.0795	.0787	.0797	.0782	.0756	.076	.0764
6	.078	.078	.079	.0753	.0734	.074	.0746
7	.0766	.0776	.0787	.0737	.0712	.072	.0728
8	.0762	.0772	.0784	.072	.069	.070	.071
9	.0758	.0769	.078	.070	.0668	.068	.0692
10	.0754	.0766	.0776	.068	.0646	.066	.0674
11	.0751	.0763	.0774	.066	.0626	.064	.0656
12	.0748	.0759	.077	.064	.0602	.062	.0638
13		.0755	.0766	.062	.058	.060	.062

	JM	L	L11	L12	LS
1	.089	.089	.089	.089	.088
2	.085	.085	.084	.085	.085
3	.081	.081	.0801	.081	.080
4	.078	.079	.0772	.078	.077
5	.0752	.077	.0745	.0755	.0735
6	.0729	.075	.0722	.073	.0705

		JM		L	L11	L12	LS
7		.0702		.0735	.0702	.0702	.068
8		.068		.072	.0675	.0675	.0658
9		.0653		.071	.065	.065	.0636
10		.0627		.070	.0625	.0625	.0613
11		.060		.069	.060	.060	.059
12		.059		.068	.0575	.0575	.057
13		.058			.055	.055	.055

	LS1		M	M1	M2	M5	M6
1	.088		.089	.089	.089	.089	.089
2	.085		.0855	.085	.085	.085	.085
3	.080		.0832	.0817	.0817	.0817	.0817
4	.0765		.0808	.0796	.080	.0792	.0801
5	.073		.0785	.0777	.0785	.0776	.0786
6	.071		.0763	.0757	.0765	.076	.077
7	.0698		.0745	.0737	.0745	.0745	.0754
8	.0678		.0726	.0717	.0725	.0729	.0738
9	.066		.0707	.0698	.0705	.0712	.0723
10	.064		.0688	.0678	.0688	.0696	.0707
11	.062		.0669	.0659	.0669	.068	.0691
12	.060		.065	.064	.0665	.0666	.0676
13	.058						

	M7	M8	M9	MA	MB	MME	MO
1	.089	.089	.089	.089	.089	.089	.089
2	.085	.086	.085	.085	.085	.085	.0855
3	.0822	.0827	.0827	.0817	.0815	.0813	.0835
4	.0807	.0812	.081	.0795	.079	.078	.0815
5	.0792	.0797	.0792	.078	.0767	.074	.080
6	.0777	.0782	.0777	.0765	.074	.0707	.07875
7	.0762	.0767	.0762	.075	.0715	.0673	.0775
8	.0747	.0752	.0747	.0735	.068	.0636	.07625
9	.0732	.0737	.0732	.072	.0653	.060	.075
10	.0717	.0722	.0717	.0705	.0627	.0563	.07375
11	.0702	.0707	.0702	.069	.060	.053	.0725
12	.0687	.0692	.0687	.0675	.059	.0495	.07125
13					.058	.046	

	MOW	MW		1	2	3	4
1	.089	.089		.089	.089	.089	.089
2	.0855	.0855		.085	.085	.085	.085
3	.0837	.084		.0814	.0814	.0814	.0814
4	.082	.0825		.0785	.0785	.0785	.0785
5	.0808	.0813		.077	.0767	.0765	.0761
6	.0795	.0803		.0755	.0749	.0744	.0737
7	.0783	.0792		.074	.0732	.0723	.0714
8	.077	.078		.0725	.0714	.0703	.0692
9	.0758	.0768		.071	.0696	.0683	.0668
10	.0745	.0757		.0695	.0678	.0661	.0645
11	.0732	.0746		.068	.066	.064	.062
12	.072	.0735		.067	.065	.063	.061
13				.066	.064	.062	.060

	5	6	7	20	21	24	24A
1	.089	.089	.089	.089	.089	.088	.088
2	.085	.085	.085	.0855	.0855	.0845	.0845
3	.0814	.0814	.0814	.0822	.0827	.0815	.0815
4	.0785	.0785	.0785	.0792	.080	.0785	.0785
5	.0758	.0755	.0755	.076	.0775	.0755	.076
6	.0733	.0725	.072	.073	.075	.0725	.0738
7	.0705	.0696	.0686	.0703	.0723	.0702	.0716
8	.068	.0666	.065	.0677	.0697	.0682	.0698
9	.0653	.0636	.062	.0647	.0671	.0667	.068
10	.0627	.0606	.0587	.0624	.0646	.065	.066
11	.060	.0577	.0553	.060	.0628	.0636	.064
12	.059	.055	.052	.058	.0608	.0625	.062
13	.058	.052	.049	.056	.059	.061	.060

	24B	61	62	69	80	81
1	.088	.089	.089	.089	.089	.089
2	.0845	.085	.085	.085	.085	.085
3	.0815	.0805	.081	.0805	.080	.080
4	.0795	.0775	.078	.0785	.078	.077
5	.0775	.0745	.075	.0765	.0758	.0742
6	.0755	.0725	.0734	.075	.0733	.0719
7	.0735	.0709	.0719	.074	.0705	.0699
8	.0715	.069	.070	.073	.068	.068
9	.0695	.067	.0685	.072	.0653	.0653
10	.0675	.0653	.067	.071	.0627	.0627
11	.0655	.0635	.065	.070	.060	.060
12	.0635	.062	.0635	.069	.059	.059
13	.0615	.060	.062		.058	.058

	O7		P4	P6	P61	PJ
1	.089		.088	.088	.088	.089
2	.085		.0845	.0845	.0845	.085
3	.0815		.081	.081	.081	.0822
4	.0795		.0785	.0775	.077	.0807
5	.078		.0755	.0738	.0746	.0795
6	.077		.073	.0703	.0715	.079
7	.076		.070	.067	.0684	.0785
8	.075		.0675	.0638	.0653	.078
9	.074		.0645	.0603	.062	.0775
10	.073		.06175	.057	.059	.077
11	.072		.059	.054	.056	.0765
12	.071		.0562	.0505	.053	.076
13			.0535	.0473	.050	

	QA	QW		R3	R6	R32	RLB
1	.089	.089		.089	.089	.088	.089
2	.085	.085		.085	.085	.084	.085
3	.082	.0825		.0805	.081	.0805	.081
4	.0796	.0802		.0776	.077	.0776	.077
5	.077	.0780		.0759	.0732	.0759	.0732
6	.0745	.0755		.074	.070	.074	.0693
7	.072	.0731		.0722	.0665	.0722	.066
8	.0695	.0708		.0703	.063	.0703	.0626
9	.0668	.0685		.0687	.060	.0687	.0592
10	.0642	.066		.067	.057	.067	.056
11	.0615	.0636		.065	.054	.065	.0526
12	.059	.0613		.0632	.051	.0632	.0495
13	.0565	.059			.048		.046

	RLS	RO	RS		S	S4	S5
1	.089	.089	.089		.089	.0895	.0895
2	.085	.085	.085		.085	.0852	.0852
3	.0804	.081	.0823		.0822	.082	.0815
4	.076	.077	.081		.0806	.079	.078
5	.0725	.074	.0803		.079	.0765	.0753
6	.0692	.071	.0795		.0774	.0744	.073
7	.066	.0685	.0786		.0758	.0722	.0707
8	.0626	.066	.078		.074	.070	.0684
9	.0592	.0633	.0775		.0724	.0677	.0663
10	.056	.0605	.077		.0708	.0656	.0639
11	.0526	.058	.0765		.069	.0634	.0616
12	.0495	.0558	.076		.0675	.0612	.0593
13	.046	.053				.059	.057

	S6		TB		V2	V3	VS
1	.089		.089		.089	.089	.089
2	.085		.085		.085	.085	.085
3	.0805		.081		.082	.0826	.081
4	.077		.078		.0795	.0804	.077
5	.0743		.075		.0775	.0783	.0748
6	.0715		.073		.0756	.0764	.0725
7	.069		.071		.0738	.0746	.071
8	.0666		.0698		.0718	.0726	.0695
9	.064		.0678		.0698	.0706	.068
10	.062		.066		.0678	.0686	.0665
11	.059		.064		.0658	.0666	.065
12	.0568		.062		.064	.0646	.0635
13	.0545		.060				

		W3	WX	WX1
1		.088	.089	.089
2		.083	.085	.085
3		.0805	.0814	.082
4		.078	.0785	.0795
5		.0763	.0775	.0777
6		.0745	.0765	.0765
7		.073	.0755	.0755
8		.071	.0746	.0746
9		.0694	.0736	.0736
10		.0677	.0728	.0728
11		.066	.0718	.0718
12		.065	.0709	.0709

.100 Jet Needles

	A9	BAA*	BAB*	BAC*	BAD*	BAE*	BAF*
1	.098	.099	.099	.099	.099	.099	.099
2	.0946	.095	.095	.095	.095	.095	.095
3	.0913	.0925	.0924	.0932	.0928	.0915	.093
4	.088	.0895	.0897	.0907	.090	.0885	.0905
5	.085	.087	.0876	.0875	.0873	.0860	.0875
6	.0834	.085	.0858	.0852	.0852	.0840	.0832
7	.0818	.0823	.084	.0823	.083	.0815	.080
8	.0802	.0792	.0822	.0763	.0808	.0790	.0768
9	.0787	.076	.0803	.0703	.0782	.0768	.0738
10	.077	.0729	.0784	.0642	.0755	.0743	.0709
11	.0755	.0697	.0759	.058	.073	.0729	.0677
12	.074	.0665	.0734	.052	.0702	.0715	.0646
13	.0722	.0633	.071	.046	.0675	.0700	.0616
14	.0706	.060	.069	.040	.065	.0685	.0584
15		.0567	.067	.040	.0624	.0670	.0554
16		.0534	.065	.040	.0598	.0655	.0523

	BAG*	BAH*	BAJ*	BAK*	BAL*	BAM*	BAN*
1	.099	.0995	.099	.099	.0972	.099	.099
2	.095	.0967	.095	.095	.0957	.095	.095
3	.0928	.0939	.0932	.0932	.0926	.0915	.0925
4	.090	.0909	.0905	.0907	.0898	.088	.0905
5	.0873	.0881	.0877	.0875	.0870	.0848	.0882
6	.0845	.0848	.0845	.0852	.0826	.0821	.0853
7	.081	.0781	.081	.0823	.0787	.0796	.082
8	.0782	.074	.0782	.0792	.0752	.0773	.079
9	.0758	.0703	.0758	.076	.0727	.0750	.0755
10	.0735	.0677	.0735	.0729	.0703	.0730	.0725
11	.0713	.0661	.0713	.0697	.0678	.0713	.0698
12	.069	.0649	.069	.0665	.0653	.0692	.0668
13	.067	.0636	.067	.0633	.0629	.0672	.0638
14	.0648	.0624	.0648	.060	.0605	.0650	.0608
15	.0625	.0618	.0625	.0568	.0580	.0630	.0578
16	.0603	.0618	.0603	.0536	.0555	.0610	.0548

	BAP*	BAQ*	BAR*	BAS*	BAT*	BAU*	BAV*
1	.099	.099	.099	.099	.099	.099	.0995
2	.0962	.095	.095	.095	.095	.095	.0967
3	.0934	.093	.0915	.0925	.0918	.0925	.0939
4	.0899	.0905	.0895	.0905	.0887	.0893	.0909
5	.0861	.088	.0877	.088	.086	.0862	.0881
6	.0826	.0845	.0853	.0845	.084	.083	.0851
7	.0791	.0810	.0820	.081	.0815	.080	.080
8	.0757	.0787	.079	.0787	.0790	.078	.0757
9	.0723	.0767	.0755	.0767	.0768	.077	.0727
10	.0688	.0745	.0725	.0745	.0743	.0764	.070
11	.0653	.0725	.0698	.0725	.0729	.0758	.0682
12	.0618	.0705	.0668	.0705	.0715	.0752	.0667
13	.0584	.0685	.0638	.0685	.070	.0747	.0655
14	.0550	.0665	.0608	.0665	.0685	.0740	.0645
15	.0516	.0645	.0578	.0645	.067	.0734	.0635
16	.0482	.0625	.0548	.0625	.0655	.0728	.0625

	BAW*	BAX*	BAY*	BAZ*	BBA*	BBB*	BBC*
1	.099	.099	.099	.099	.099	.099	.099
2	.095	.0958	.0956	.0952	.0957	.0956	.095
3	.0918	.0926	.0926	.0926	.0934	.0926	.093
4	.0887	.0896	.0895	.0896	.0899	.0896	.0902
5	.086	.0861	.0858	.0866	.0861	.0865	.0873
6	.0827	.0814	.0815	.0820	.0826	.0815	.0848
7	.0799	.0774	.0774	.0781	.0791	.0774	.0820
8	.0774	.0742	.0742	.0751	.0757	.0742	.0752
9	.0755	.0718	.0718	.0727	.0723	.0718	.0687
10	.0735	.0694	.0694	.0703	.0688	.0694	.0620
11	.0715	.0673	.0673	.0678	.0653	.0673	.0554
12	.070	.0652	.0652	.0654	.0618	.0652	.0487
13	.069	.063	.063	.0629	.0584	.063	.042
14	.068	.061	.061	.0605	.0550	.061	.042
15	.067	.059	.059	.058	.0516	.059	.042
16	.065	.057	.057	.056	.0482	.057	.042

	BBD*	BBE*	BBF*	BBG*	BBH*	BBK*	BBL*
1	.099	.099	.099	.099	.099	.099	.099
2	.095	.095	.095	.095	.095	.095	.095
3	.092	.093	.0929	.0932	.092	.092	.0925
4	.090	.0905	.0905	.0905	.090	.089	.0893
5	.088	.0882	.0877	.0878	.0879	.086	.0862
6	.0862	.085	.0845	.0852	.0853	.0827	.083
7	.0844	.082	.0813	.0829	.082	.0799	.0825
8	.0825	.078	.0795	.0806	.079	.0780	.082
9	.0818	.0757	.0777	.0783	.0755	.0767	.0818
10	.0608	.0733	.0761	.076	.0725	.0753	.0808
11	.0798	.071	.0744	.0737	.0698	.0740	.0798
12	.0788	.069	.072	.0713	.0668	.0727	.0788
13	.0778	.067	.0695	.069	.0638	.0713	.0778
14	.0768	.065	.067	.0668	.0608	.0700	.0768
15	.0758	.063	.0647	.0644	.0578	.0687	.0758
16	.0748	.061	.0623	.0622	.0548	.0675	.0748

	BBM*	BBN*	BBP*	BBQ*	BBR*	BBS*	BBT*
1	.099	.099	.099	.099	.099	.099	.099
2	.095	.095	.095	.095	.095	.095	.096
3	.093	.0925	.0912	.093	.0926	.0915	.0932
4	.0905	.0905	.0886	.0908	.0902	.0885	.0903
5	.087	.0874	.0859	.0877	.0881	.0870	.0877
6	.0843	.0862	.0832	.0852	.0861	.0850	.0850
7	.0795	.0846	.0803	.0813	.0840	.0823	.0827
8	.075	.083	.0775	.0795	.0825	.0806	.0807
9	.074	.082	.0752	.0777	.0810	.0792	.0792
10	.073	.0805	.0727	.076	.0788	.0777	.0778
11	.071	.080	.0710	.0744	.0770	.0763	.0765
12	.069	.079	.0689	.072	.0750	.0748	.0753
13	.067	.078	.0669	.0696	.0730	.0734	.0740
14	.065	.077	.0647	.067	.0710	.0719	.0725
15	.063	.076	.0627	.0647	.0690	.0709	.0713
16	.061	.075	.0607	.0623	.0670	.0689	.0700

	BBU*	BBV*	BBW*	BBX*	BBY*	BBZ*
1	.099	.099	.099	.099	.099	.098
2	.095	.095	.095	.095	.095	.0954
3	.0932	.0932	.0923	.0932	.0915	.0924
4	.0907	.0907	.090	.0905	.0882	.0892
5	.0876	.0875	.0870	.0875	.0867	.0862
6	.0859	.0852	.0832	.0852	.0848	.0819
7	.0840	.0829	.0792	.0829	.0821	.0780
8	.0822	.0805	.0750	.0806	.0803	.0751
9	.0805	.0773	.0717	.0782	.0790	.0713
10	.0788	.0742	.0682	.0755	.0775	.0678
11	.0770	.0710	.0647	.0730	.0761	.0653
12	.0752	.0679	.0610	.0702	.0746	.0629
13	.0734	.0648	.0577	.0675	.0732	.0605
14	.0718	.0617	.0540	.0650	.0717	.0580
15	.0700	.0595	.0505	.0624	.0707	.0560
16	.0680	.0552	.0470	.0598	.0687	.0540

	BCA*	BCB*	BCC*	BCD*	BCE*	BCF	BCG*
1	.099	.099	.099	.1000	.099		.099
2	.0955	.095	.095	.0960	.095		.095
3	.093	.0912	.0924	.0930	.090		.0914
4	.0897	.0886	.0897	.0902	.0872		.0886
5	.0854	.0859	.0855	.0872	.0835		.0837
6	.0795	.0841	.0820	.0843	.0795		.0786
7	.073	.0823	.0800	.0814	.0768		.0736
8	.0676	.0806	.079	.0786	.0740		.0665
9	.0639	.0792	.0775	.0757	.0705		.0582
10	.0575	.0777	.077	.0727	.0675		.0493
11	.0519	.0763	.0759	.0700	.0645		.0414
12	.0498	.0748	.0734	.067	.0615		.0331
13	.045	.0734	.0710	.064	.0585		.0331
14	.040	.0719	.0690	.061	.0555		.0331
15	.040	.0709	.0670	.058	.0525		.0331
16	.040	.0689	.0650	.055	.0495		.0331

	BCH*	BCJ*	BCK*	BCL	BCM*	BCN*	BCP*
1	.099	.0995	.099		.099	.099	.099
2	.095	.0967	.0953		.096	.096	.095
3	.092	.0939	.0922		.0932	.0932	.0929
4	.0892	.0909	.0903		.0905	.0905	.0905
5	.0863	.0881	.0881		.0882	.088	.0881
6	.0833	.0848	.0832		.0860	.0857	.0858
7	.0800	.0781	.080		.0842	.0837	.0835
8	.0772	.0740	.0768		.0827	.082	.0804
9	.0745	.0703	.0738		.0815	.0805	.0793
10	.0722	.0671	.0709		.0806	.0724	.0781
11	.0700	.0650	.0677		.0797	.0786	.0759

	* BCH	* BCJ	* BCK	* BCL	* BCM	* BCN	* BCP
12	.0678	.0630	.0646		.0790	.078	.0737
13	.0656	.0610	.0616		.0785	.0775	.0715
14	.0634	.0600	.0584		.0780	.077	.0693
15	.0612	.0590	.0554		.0775	.0765	.0671
16	.0592	.0580	.0523		.0770	.076	.0649

	BC	C1	CIW	CV	DF	DG	DX
1	.099	.099	.099	.099	.100	.100	.099
2	.095	.095	.0955	.094	.098-1/16	.098-1/16	.0958
3	.091	.0916	.093	.090	.096	.096	.0926
4	.088	.0889	.0905	.0867	.091	.0905	.0888
5	.085	.0861	.0875	.084	.088	.087	.0859
6	.0825	.084	.0856	.0815	.085	.0836	.0830
7	.0803	.0818	.0836	.0795	.082	.0804	.0775
8	.0785	.0796	.0819	.0775	.0782	.0772	.0740
9	.0765	.0778	.080	.076	.075	.074	.0705
10	.075	.076	.078	.0745	.0725	.071	.0670
11	.073	.074	.076	.0732	.070	.0689	.0635
12	.072	.073	.0752	.072	.0685	.067	.0600
13	.071	.072	.0746	.071	.067	.0653	.0565
14	.070	.071	.074	.070	.0655	.0636	.0530
15					.064	.062	

	GK		K	KA	KB	KC	KD
1	.100		.100	.099	.099	.099	.099
2	.094		.095	.095	.095	.0935	.095
3	.089		.0905	.092	.0917	.090	.0908
4	.085		.0865	.0895	.090	.087	.0883
5	.081		.083	.0876	.0886	.084	.0856
6	.078		.080	.0855	.0972	.082	.083
7	.075		.0777	.0835	.086	.080	.0775
8	.072		.076	.0814	.085	.0785	.074
9	.069		.074	.0793	.0845	.0765	.0705
10	.066		.0722	.0784	.084	.075	.067
11	.063		.0705	.078	.0835	.073	.0635
12	.060		.0685	.078	.083	.0715	.060
13	.057		.0667	.078	.0825	.0695	.0565
14	.054		.065	.078		.0675	.053

	KE	KF	KG	KH	KI	KK	KL
1	.099	.099	.099	.099	.098	.099	.099
2	.095	.095	.095	.0935	.094	.095	.095
3	.0917	.091	.0915	.0883	.090	.0925	.0925
4	.090	.089	.0895	.0863	.0865	.0893	.0895
5	.0885	.0875	.0875	.0843	.083	.086	.087
6	.087	.0855	.0855	.0825	.080	.0828	.085
7	.0848	.0835	.0841	.0805	.0777	.0795	.0823
8	.0838	.082	.0834	.0785	.076	.0738	.0792
9	.083	.081	.0827	.076	.074	.0615	.076
10	.0822	.080	.0820	.0735	.0722	.0517	.0729
11	.0815	.0795	.0810	.071	.0705	.042	.0697
12	.0808	.079	.0800	.0685	.0685	.042	.0665
13	.080	.079	.0790	.066	.0667	.042	.0633
14	.0792	.079	.0780		.065	.042	.060

	KM	* KN	* KN cont.	KO	KP	KQ	KR
1	.099	.099	.066	.099	.099	.099	.099
2	.095	.0962	.063	.095	.095	.095	.095
3	.0924	.0933	.060	.0925	.092	.093	.0932
4	.0897	.0905	.057	.0895	.0893	.0903	.0907
5	.0876	.087	.054	.087	.0867	.088	.0875
6	.0858	.082	.051	.085	.0842	.0855	.0852
7	.084	.0782		.0825	.0817	.082	.0823
8	.0822	.0747		.0795	.078	.0795	.077
9	.0803	.0718		.0765	.070	.0775	.071
10	.0784	.069		.074	.0635	.0755	.0658
11	.0759			.0715	.058	.0735	.0603
12	.0734			.069	.0523	.0715	.057
13	.071			.067	.0465	.0695	.0534
14	.069			.065	.0418	.0675	.050

	KS	KT	KTA	KU	KV	KW	KW1
1	.099	.100	.100	.099	.099	.100	.100
2	.095	.095	.095	.095	.095	.095	.095
3	.092	.0905	.0895	.0927	.0913	.091	.0915
4	.0892	.087	.0865	.0897	.088	.0877	.0885
5	.0875	.0845	.0835	.0868	.0838	.084	.0851
6	.085	.0822	.0815	.084	.080	.0805	.082
7	.079	.0802	.0802	.0812	.0736	.077	.079
8	.075	.079	.0788	.0783	.0650	.073	.076
9	.070	.0775	.0775	.0756	.058	.0695	.0728
10	.065	.076	.076	.0729	.0506	.0661	.0695
11	.060	.075	.075	.070	.0435	.063	.0663
12	.055	.074	.074	.0673	.0363	.060	.063
13	.050	.073	.073	.0645	.0363	.057	.060
14	.045	.072	.072	.0618	.0363	.054	.057

	KW2	KWR	LB1	LB2	LFN	LBA
1	.100	.100	.100	.100	.099	.100
2	.095	.095	.096	.095	.094	.096
3	.095	.090	.0925	.091	.0885	.093
4	.0875	.087	.0895	.0895	.085	.0906
5	.0845	.084	.087	.086	.0815	.0883
6	.0815	.081	.085	.084	.078	.0863
7	.0788	.0786	.083	.0825	.075	.0848
8	.076	.076	.082	.0815	.0723	.0839.
9	.0728	.0728	.081	.0805	.0695	.0830
10	.0695	.0695	.080	.0795	.0668	.0821
11	.0663	.0663	.079	.0785	.064	.0812
12	.063	.063	.078	.0775	.061	.0804
13	.060	.060	.077	.0765	.0585	.0796
14	.057	.057	.076	.0755	.0556	.0788

	MC	53	55	58
1	.100	.100	.100	.100
2	.095	.095	.095	.095
3	.0915	.090	.090	.090
4	.0885	.0858	.0858	.0858
5	.0851	.0815	.0815	.0815
6	.082	.0777	.078	.078
7	.079	.0738	.074	.0744
8	.076	.070	.070	.0712
9	.0728	.0668	.067	.0685
10	.0684	.0636	.064	.066
11	.064	.0606	.062	.064
12	.060	.0579	.060	.062
13	.0555	.0552	.057	.060
14	.051	.053	.055	.058

	OA6	OA7	OA8
1	.100	.100	.100
2	.096	.096	.097
3	.0917	.093	.094
4	.0887	.0902	.0913
5	.0856	.0872	.0888
6	.0825	.0843	.086
7	.0794	.0814	.0835
8	.0762	.0786	.0808
9	.0731	.0757	.0782
10	.070	.0727	.0755
11	.067	.070	.073
12	.064	.067	.0702
13	.061	.064	.0675
14	.058	.061	.065

	RA	RB	RC	RD	RE	RF	RG
1	.100	.099	.099	.099	.099	.100	.100
2	.094	.095	.0946	.095	.095	.095	.095
3	.088	.0907	.090	.090	.092	.090	.0905
4	.083	.0866	.0855	.0865	.089	.0863	.087
5	.0785	.0825	.081	.0835	.0856	.0825	.0836
6	.0753	.0784	.0765	.081	.0805	.0788	.0802
7	.0722	.074	.072	.079	.0753	.075	.0768
8	.069	.070	.0674	.077	.0712	.0712	.0732
9	.0658	.0657	.0627	.0757	.0670	.0675	.070
10	.0627	.0615	.0583	.0742	.0628	.0637	.0665
11	.0595	.0575	.0537	.073	.0587	.060	.0630
12	.0564	.0532	.0492	.0715	.0543	.057	.060
13	.053	.049	.0446	.0703	.0501	.054	.057
14	.050	.045	.040	.069	.0460	.051	.054

	RH	RI	RJ	RK	RL	RN	RP
1	.100	.099	.100	.100	.100	.099	.099
2	.095	.095	.095	.095	.095	.095	.095
3	.0915	.0912	.0910	.0915	.0920	.092	.0905
4	.0878	.0881	.087	.088	.089	.089	.087
5	.0846	.0846	.0844	.0854	.086	.0859	.085

	RH	RI	RJ	RK	RL	RN	RP
6	.0813	.0804	.0818	.0830	.0835	.0808	.0832
7	.078	.0763	.0792	.0805	.0810	.076	.082
8	.075	.0722	.0766	.078	.0785	.0721	.0815
9	.072	.067	.0740	.0756	.0760	.0688	.0809
10	.069	.0605	.0714	.0732	.0735	.0651	.080
11	.066	.0567	.0688	.0708	.0710	.0617	.0794
12	.063	.0525	.0662	.0683	.0685	.0581	.0786
13	.060	.049	.0630	.066	.0660	.0547	.0778
14	.057	.045	.0610	.0635	.0636	.0510	.077

	RR	RU	RV			SA	SB
1	.099	.100	.100			.099	.099
2	.095	.095	.095			.095	.095
3	.0922	.089	.089			.0915	.091
4	.0895	.084	.084			.0885	.0875
5	.086	.081	.081			.086	.084
6	.0808	.0785	.0785			.084	.080
7	.076	.0756	.0755			.0815	.076
8	.0721	.0733	.073			.079	.072
9	.0688	.071	.070			.0768	.0695
10	.0651	.0688	.067			.0743	.067
11	.0617	.066	.0645			.072	.064
12	.0581	.0636	.0618			.070	.062
13	.0547	.061	.059			.068	.059
14	.051	.059	.056			.066	.057

	SC	SD	SE	SF	SG	SH	SJ
1	.100	.100	.100	.100	.100	.099	.099
2	.095	.095	.095	.095	.095	.0952	.095
3	.0915	.0912	.0915	.09125	.0917	.0917	.092
4	.0882	.088	.0903	.0895	.090	.0885	.0895
5	.0865	.0862	.0888	.08825	.0887	.0865	.0875
6	.0845	.0842	.087	.087	.0878	.0845	.0855
7	.0823	.0822	.0852	.08575	.0869	.082	.0835
8	.080	.080	.0845	.085	.0862	.0793	.0815
9	.079	.079	.084	.0845	.0859	.077	.0792
10	.078	.078	.084	.084	.0855	.0743	.077
11	.077	.077	.084	.084	.0855	.0722	.075
12	.076	.076	.084	.084	.0855	.070	.073
13	.075	.075	.084	.084	.0855	.068	.071
14	.074	.074	.084	.084	.0855	.066	.069

	SK	SL	SM	SN	SO	SP	SQ
1	.099	.099	.099	.099	.100	.099	.100
2	.0954	.095	.095	.095	.096	.095	.095
3	.093	.092	.0915	.0915	.092	.0915	.0915
4	.090	.089	.0885	.089	.089	.0892	.088
5	.0867	.0867	.0855	.0865	.086	.0871	.0854
6	.0833	.0833	.0825	.0845	.083	.0852	.083
7	.080	.080	.0795	.0835	.0804	.0834	.0812
8	.077	.077	.0765	.082	.0778	.0810	.0794
9	.074	.074	.0735	.080	.075	.0786	.0775
10	.0714	.0714	.071	.078	.0743	.0762	.0757
11	.0695	.0695	.069	.0765	.0737	.0738	.0738
12	.0676	.-676	.067	.075	.073	.0714	.0719
13	.065	.065	.065	.074	.0722	.069	.070
14		.063	.063	.073		.0666	.068

	SR	SS	ST	SU	SV	SW	SX
1	.099	.099	.099	.099	.099	.099	.099
2	.095	.0945	.095	.095	.0955	.0955	.095
3	.092	.0905	.0925	.0902	.0923	.0925	.0902
4	.089	.086	.0895	.0868	.089	.088	.0868
5	.086	.084	.087	.0834	.0867	.086	.084
6	.083	.080	.0852	.0802	.0845	.084	.0808
7	.0804	.0755	.0831	.0764	.0818	.082	.0774
8	.0778	.0715	.0805	.0728	.079	.0795	.0744
9	.075	.0655	.0787	.069	.0755	.077	.0713
10	.0743	.059	.077	.0655	.072	.0745	.0686
11	.0737	.0535	.0753	.062	.0685	.072	.0657
12	.073	.0485	.0737	.0586	.0665	.0695	.063
13	.0722	.045	.071	.0552	.0653	.067	.060
14	.0715	.042	.069	.0518	.064	.0645	.0572

	SY	SZ			TA	TC	TD
1	.099	.099			.099	.099	.099
2	.095	.0945			.095	.095	.095
3	.091	.0903			.0915	.0915	.0925
4	.0875	.0858			.0882	.089	.090
5	.0855	.0837			.0867	.087	.0875
6	.0825	.0798			.0852	.085	.0845
7	.078	.0753			.0831	.0832	.0803
8	.075	.0707			.0805	.0825	.077
9	.069	.0643			.0787	.0815	.073
10	.063	.0570			.0770	.0805	.0678
11	.0575	.0508			.0753	.0795	.0631
12	.0525	.0470			.0737	.0785	.0585
13	.049	.0445			.0710	.0775	.0538
14	.046	.0420			.0690	.0765	.049

	TE	TF	TG	TH	TJ	TK	TL
1	.099	.099	.099	.099	.099	.099	.099
2	.095	.096	.095	.092	.095	.095	.095
3	.092	.0925	.0922	.0908	.092	.092	.092
4	.0895	.089	.0895	.0893	.0895	.090	.089
5	.087	.0855	.0878	.0878	.0877	.0885	.086
6	.0837	.082	.0862	.0863	.086	.087	.0835
7	.0805	.0782	.0847	.0848	.0845	.086	.081
8	.076	.0748	.0837	.0825	.084	.0855	.0793
9	.0715	.0714	.0827	.074	.083	.0845	.0776
10	.067	.068	.0818	.065	.082	.0835	.0759
11	.0625	.0645	.0812	.055	.081	.0825	.0746
12	.058	.061	.0806	.049	.080	.0815	.0733
13	.0535	.0575	.0800	.044	.079	.0805	.072
14	.049	.054	.0794	.040	.078	.0795	.071

	TM	TN	TO	TP	TR	TS	TT
1	.099	.099	.099	.099	.099	.099	.0989
2	.095	.0945	.095	.0955	.095	.095	.0941
3	.092	.091	.092	.092	.092	.0915	.0904
4	.0894	.087	.0893	.090	.089	.088	.0883
5	.0867	.085	.0865	.0875	.0869	.0853	.0861
6	.0842	.082	.0845	.0856	.084	.0833	.0825
7	.0814	.078	.083	.0836	.0817	.0812	.0799
8	.0785	.073	.082	.0819	.0791	.0782	.078
9	.0775	.0665	.081	.080	.0765	.0742	.0766
10	.077	.060	.080	.078	.0738	.0710	.0748
11	.077	.0535	.079	.076	.071	.0675	.0733
12	.077	.050	.078	.0752	.0685	.061	.0721
13	.077	.0475	.077	.0746	.063	.055	.0715
14	.077	.045	.076	.074	.062	.049	

	TU	TV	TW	TX	TY	TZ
1	.099	.099	.099	.099	.099	.099
2	.095	.095	.095	.095	.095	.095
3	.091	.0925	.0925	.091	.0918	.0915
4	.088	.090	.0895	.0894	.0887	.0893
5	.085	.088	.087	.0867	.086	.087
6	.083	.086	.0852	.0842	.0836	.0847
7	.081	.084	.0831	.0814	.0825	.0827
8	.0793	.0825	.0805	.0785	.0805	.0805
9	.0776	.081	.0775	.078	.0775	.0787
10	.0759	.080	.075	.078	.075	.077
11	.0746	.0795	.0722	.078	.0722	.0753
12	.0733	.0785	.070	.078	.070	.0737
13	.072	.0775	.068	.078	.068	.071
14	.071	.0765	.066	.078	.066	.069

		WO2	WO3	WO4		ZA	ZB
1		.100	.100	.100		.099	.099
2		.095	.095	.095		.095	.095
3		.091	.091	.090		.0905	.0915
4		.087	.08775	.086		.0875	.089
5		.0835	.0845	.082		.0847	.0865
6		.081	.0822	.0794		.0822	.084
7		.0785	.080	.0768		.080	.082
8		.076	.078	.074		.0778	.0795
9		.0732	.0755	.0712		.0758	.0775
10		.071	.0735	.069		.0737	.0755
11		.0683	.0712	.066		.071	.0735
12		.0657	.069	.0634		.0675	.071
13		.063	.067	.061		.063	.067
14		.061	.065	.058		.0585	.063
15						.054	.059

	ZC	ZD	ZE	ZF	ZG	ZH
1	.099	.099	.099	.099	.099	.099
2	.095	.095	.095	.095	.095	.095
3	.091	.090	.0895	.089	.089	.0915
4	.088	.0865	.0858	.085	.0845	.0893
5	.0855	.084	.083	.0821	.0815	.087
6	.083	.0815	.0805	.0798	.079	.0847
7	.081	.079	.0782	.0775	.0765	.082
8	.0785	.077	.076	.075	.074	.0795
9	.0765	.0745	.0732	.072	.071	.0775

	ZC	ZD	ZE	ZF	ZG	ZH
10	.0745	.072	.0703	.0687	.0675	.0755
11	.0725	.069	.0673	.065	.063	.0735
12	.0695	.065	.0629	.0608	.059	.0715
13	.065	.0605	.0584	.0563	.054	.0695
14	.0605	.056	.054	.052	.050	.0675
15	.056	.0515	.0495	.0475	.045	

.125 Jet Needles

	NA	UA	UB	UC	UD	UE
1	.124	.124	.124	.124	.124	.124
2	.1205	.1205	.1208	.1205	.1205	.1205
3	.1170	.1175	.1165	.118	.1178	.1155
4	.1144	.1146	.113	.1153	.1158	.1135
5	.1125	.1117	.111	.1128	.114	.1112
6	.1108	.1074	.109	.1107	.1126	.109
7	.1090	.1023	.107	.1086	.1115	.107
8	.1078	.0974	.1055	.107	.1104	.105
9	.1068	.0931	.104	.1056	.1092	.103
10	.1065	.0890	.103	.1046	.108	.1015
11	.1065	.0849	.102	.104	.1069	.100
12	.1065	.0808	.101	.1032	.1064	.099
13	.1065	.0767	.100	.1025	.1058	.098
14	.1065	.0726	.099	.1018	.1047	.0965
15	.1065	.0685	.098	.101	.1036	.095
16	.1065	.0644	.097	.1002	.1025	.0935

	UF	UG	UH	UI	UJ	UK	UL
1	.124	.124	.124	.124	.124	.124	.124
2	.1205	.1205	.1205	.1205	.1205	.1205	.1205
3	.1163	.116	.1165	.1172	.1165	.116	.1173
4	.113	.112	.1135	.114	.1135	.113	.1140
5	.1106	.1084	.1105	.1114	.1105	.1106	.1113
6	.1073	.1054	.1085	.1092	.108	.1073	.1090
7	.1035	.1023	.1065	.107	.1055	.1035	.1076
8	.0997	.099	.104	.1051	.103	.0997	.106
9	.0960	.0954	.1015	.1032	.100	.0960	.1035
10	.0926	.0923	.099	.1018	.097	.0927	.1006
11	.089	.0895	.0965	.1006	.094	.090	.098
12	.0854	.0875	.094	.0995	.091	.0875	.0956
13	.0819	.086	.0915	.0895	.0875	.0860	.0937
14	.0783	.0845	.089	.0975	.0845	.0845	.0917
15	.0748	.083	.0865	.0965	.0815	.0830	.0897
16	.0713	.0815	.084	.0955	.080	.0815	.0877

	UM	UN	UO	UP	UR	US	UT
1	.124	.124	.124	.124	.124	.124	.124
2	.1205	.1205	.1205	.1205	.1205	.1205	.1205
3	.1165	.1165	.1155	.116	.116	.1172	.1147
4	.114	.113	.1135	.114	.1128	.1133	.1114
5	.1123	.110	.110	.1105	.1094	.1105	.1084
6	.1104	.107	.108	.1075	.106	.1092	.1054
7	.1086	.104	.1055	.1045	.103	.1085	.1023
8	.107	.1005	.103	.1025	.0997	.1068	.0990
9	.1056	.0985	.101	.101	.096	.1043	.0954
10	.1046	.0965	.0985	.0995	.0927	.102	.0917
11	.104	.0945	.0965	.0985	.090	.1006	.088
12	.1032	.0915	.094	.0975	.0875	.0995	.0858
13	.1025	.0885	.093	.096	.0860	.0985	.0836
14	.1018	.0855	.0915	.0945	.0945	.0975	.0813
15	.101	.0830	.090	.093	.083	.0965	.079
16	.1002	.0805	.089	.0915	.0815	.0955	.0768

	UU	UV	UW	UX	UY	UZ
1	.124	.124	.124	.124	.124	.124
2	.1205	.1205	.1205	.1205	.1205	.1205
3	.1165	.1182	.1175	.1165	.1174	.1172
4	.114	.1160	.1146	.114	.1145	.1135
5	.112	.1135	.1117	.112	.1128	.1113
6	.1095	.1112	.1074	.1095	.1107	.1095
7	.107	.1095	.1023	.107	.1086	.1084
8	.105	.108	.0974	.105	.107	.1066
9	.1025	.1065	.092	.1025	.1056	.1048
10	.100	.1053	.0865	.100	.1046	.103
11	.099	.1047	.081	.099	.104	.102
12	.098	.1042	.0755	.098	.1032	.101
13	.097	.1036	.070	.0955	.1025	.100
14	.096	.1029	.064	.0931	.1018	.099
15	.0945	.1029	.058	.091	.101	.098
16	.093	.1029	.052	.089	.1002	.097

	VA UVA	VB UVB	VC UVC	VD UVD	VE UVE	VF UVF	VG UVG
1	.124	.124	.124	.124	.124	.124	.124
2	.1188	.1188	.1188	.1184	.1178	.1172	.1166
3	.1145	.1145	.1145	.1135	.1125	.1114	.1103
4	.111	.111	.111	.1096	.108	.1063	.1047
5	.108	.108	.108	.106	.104	.102	.100
6	.1061	.1056	.1052	.103	.1008	.0985	.0962
7	.1044	.1035	.1025	.100	.0975	.095	.0925
8	.1025	.101	.0997	.097	.0943	.0915	.0887
9	.1006	.0986	.097	.094	.091	.088	.085
10	.0987	.0964	.0943	.091	.0877	.0845	.0813
11	.0968	.094	.0915	.088	.0846	.081	.0775
12	.095	.0917	.0888	.0852	.0814	.0775	.0737
13	.0932	.0895	.086	.0822	.078	.074	.070
14	.0913	.0872	.0833	.079	.075	.0715	.0662
15	.0895	.0848	.0805	.076	.0715	.067	.0625
16	.0875	.0825	.0777	.0732	.0685	.0635	.0586

	VH UVH	VI UVI	VJ UVJ	VK UVK	VL UVL	VM UVM	VN UVN
1	.124	.124	.124	.124	.124	.124	.124
2	.116	.1154	.1146	.1134	.1122	.1122	.1122
3	.109	.1077	.1065	.1046	.103	.103	.103
4	.103	.1015	.0997	.0977	.0956	.0946	.0946
5	.098	.096	.094	.092	.090	.088	.087
6	.094	.0917	.0895	.0874	.085	.0827	.080
7	.090	.0875	.085	.0825	.080	.0775	.0734
8	.086	.0833	.0805	.0778	.075	.0722	.067
9	.082	.079	.076	.073	.070	.0668	.0604
10	.078	.0748	.0715	.0684	.065	.0615	.055
11	.074	.0715	.067	.0635	.060	.0563	.0494
12	.070	.0662	.0625	.0588	.055	.051	.044
13	.066	.062	.058	.054	.050	.0457	.039
14	.062	.0577	.0535	.0493	.045	.0405	.034
15	.058	.0535	.049	.0445	.040	.035	.0295
16	.054	.0492	.0445	.0396	.035	.030	.025

	VO UVO	VP UVP	VR UVR	VT UVT	UVU	UVV	UVW
1	.124	.124	.124	.125	.1240	.124	.124
2	.121	.121	.119	.121	.1205	.1205	.1205
3	.117	.117	.1145	.116	.1172	.1165	.1157
4	.113	.113	.1108	.1115	.1135	.1146	.1107
5	.110	.110	.1075	.107	.1115	.113	.1077
6	.1075	.107	.1038	.103	.1103	.1115	.105
7	.105	.104	.1004	.099	.1091	.1098	.1012
8	.1022	.100	.097	.095	.1078	.1085	.0977
9	.0995	.0965	.0925	.091	.1062	.1071	.0942
10	.097	.093	.088	.087	.1047	.106	.0908
11	.0935	.089	.0835	.084	.1025	.1052	.0872
12	.0895	.0855	.079	.081	.1015	.1045	.0837
13	.0855	.082	.076	.0775	.1002	.1037	.0803
14	.082	.0782	.073	.0745	.0994	.1032	.0767
15	.078	.0745	.070	.072	.0984	.1025	.0732
16	.074	.0707	.067	.069	.0974	.1018	.0698
17					.0964		

	UVX	.25 U25	35 U35	45 U45	50 U50	59 U59	60 U60
1	.124	.124	.124	.124	.124	.124	.124
2	.1205	.123	.123	.123	.117	.116	.117
3	.1165	.116	.116	.116	.111	.111	.111
4	.1146	.109	.109	.109	.104	.106	.104
5	.1126	.100	.100	.101	.099	.101	.099
6	.1091	.092	.092	.094	.092	.096	.094
7	.1071	.084	.085	.089	.086	.091	.089
8	.1065	.076	.078	.082	.080	.087	.084
9	.1059	.067	.071	.076	.073	.083	.0795
10	.1056	.059	.064	.070	.068	.079	.075
11	.1050	.051	.057	.064	.062	.075	.070
12	.1045	.042	.050	.057	.056	.071	.065
13	.1037	.034	.042	.051	.050	.067	.060
14	.1032	.025	.035	.045	.044	.063	.055
15	.1025				.038	.059	.050
16	.1018						

	70 U70	74 U74	75 U75	76 U76	78 U78	79 U79	O1 UO1
1	.125	.124	.125	.124	.124	.124	.124
2	.119	.117	.119	.117	.117	.1178	.123
3	.113	.112	.113	.112	.112	.1147	.1175
4	.1085	.108	.1087	.109	.109	.1115	.1115
5	.104	.104	.105	.1055	.107	.1084	.1065
6	.0995	.100	.1001	.1022	.104	.1053	.1015

	70 U70	74 U74	75 U75	76 U76	78 U78	79 U79	01 U01
7	*.095*	*.097*	*.097*	*.099*	*.101*	*.1022*	*.097*
8	*.0905*	*.094*	*.093*	*.0958*	*.0978*	*.0991*	*.092*
9	*.086*	*.090*	*.089*	*.0924*	*.094*	*.096*	*.087*
10	*.081*	*.087*	*.085*	*.0892*	*.091*	*.093*	*.082*
11	*.077*	*.084*	*.081*	*.086*	*.088*	*.090*	*.0775*
12	*.072*	*.081*	*.077*	*.083*	*.085*	*.087*	*.0725*
13	*.068*	*.078*	*.073*	*.080*	*.082*	*.084*	*.0675*
14	*.063*	*.076*	*.069*	*.078*	*.080*	*.082*	*.063*
15	*.059*	*.074*	*.065*	*.076*	*.078*	*.080*	*.058*
16		*.072*		*.074*	*.076*	*.078*	

Printed by
J H Haynes & Co Ltd
Sparkford Nr Yeovil
Somerset BA22 7JJ England